Firebase Tan Tru

ALSO BY WALTER F. MCDERMOTT

*Understanding Combat Related
Post Traumatic Stress Disorder*
(McFarland, 2012)

Firebase Tan Tru

Memoir of an Artilleryman in the Mekong Delta, 1969–1970

WALTER F. MCDERMOTT

McFarland & Company, Inc., Publishers
Jefferson, North Carolina

LIBRARY OF CONGRESS CATALOGUING-IN-PUBLICATION DATA

Names: McDermott, Walter F., 1946– author.
Title: Firebase Tan Tru : memoir of an artilleryman in the Mekong Delta, 1969–1970 / Walter F. McDermott.
Other titles: Memoir of an artilleryman in the Mekong Delta, 1969–1970
Description: Jefferson, North Carolina : McFarland & Company, Inc., Publishers, 2018 | Includes index.
Identifiers: LCCN 2018026427 | ISBN 9781476668529 (softcover : acid free paper) ∞
Subjects: LCSH: McDermott, Walter F., 1946– | Vietnam War, 1961–1975—Campaigns—Mekong River Delta (Vietnam and Cambodia) | Artillerymen—United States—Biography. | United States. Army. Division, 9th—Biography. | Vietnam War, 1961–1975—Campaigns—Vietnam—Long An (Province) | Vietnam War, 1961–1975—Artillery operations, American. | Vietnam War, 1961–1975—Personal narratives, American.
Classification: LCC DS559.9.M45 A3 2018 | DDC 959.704/3092 [B] —dc23
LC record available at https://lccn.loc.gov/2018026427

BRITISH LIBRARY CATALOGUING DATA ARE AVAILABLE

ISBN (print) 978-1-4766-6852-9
ISBN (ebook) 978-1-4766-3292-6

© 2018 Walter F. McDermott. All rights reserved

No part of this book may be reproduced or transmitted in any form or by any means, electronic or mechanical, including photocopying or recording, or by any information storage and retrieval system, without permission in writing from the publisher.

Front cover image of Walter McDermott in the Tan Tru mud near the end of his combat tour

Printed in the United States of America

McFarland & Company, Inc., Publishers
 Box 611, Jefferson, North Carolina 28640
  www.mcfarlandpub.com

To the hundreds of combat veterans
of America's wars that I worked with
as a clinical psychologist
within the Veterans Administration.
They sacrificed much and taught me a great deal.
May they all achieve the peace of mind
they searched for.

I thought about Tolstoi and about what a great advantage an experience of war was to a writer. It was one of the major subjects and certainly one of the hardest to write truly of and those writers who had not seen it were always very jealous and tried to make it seem unimportant or abnormal, or a disease as a subject, while, really, it was just something quite irreplaceable that they had missed.

—Ernest Hemingway in *Green Hills of Africa*

Table of Contents

Preface   1

1. Avoidable Casualties   5
2. Basic Training   10
3. Advanced Individual Training   22
4. Arriving in Country   32
5. Tan An   35
6. Liaison Specialist and a Failure   44
7. Air Power Support   51
8. Dinner with a Wealthy Vietnamese Farmer   54
9. The Helicopter War   59
10. Rocket Attack   68
11. Losing the War   73
12. Time on Target Attack   76
13. Firebase Animals   79
14. Reconnaissance Platoon Tragedy   89
15. Burning Shit   96
16. Fragging   100
17. The Food   108

Table of Contents

18. Weapons	115
19. Atypical Duty	121
20. Unique Nights	125
21. Drugs	130
22. Officers	142
23. Vietnamese Enemies vs. Allies	146
24. Saigon	154
25. Tan An Sniper	160
26. Religion	165
27. Weddings	172
28. Firebase Drinking Clubs	178
29. Viet Cong Attacks	185
30. Media Biases	188
31. Coming Home	192
32. Reflections	200
Vietnam Glossary	205
Military Service History of Walter F. McDermott	207
Index	209

Preface

More than two million soldiers served in America's Vietnam War; each engendered his unique vision of the war. My memoir is a portrayal of my own perceptions and experiences fighting in the Vietnam War. America's Vietnam War was so foreign that we soldiers developed a new language to describe it. Every soldier I had any contact with, from lofty full colonels down to the lowest private, quickly learned to refer to back home in America as "the world." For example, a sergeant might state: "When I get back to the world, I am going to marry my girlfriend."

We talked this way because to all of us the Vietnam War was so alien from our previous lives that it could not be of this world. It was somewhere else, a genuine surrealistic nightmare that we could not awaken from. The many bizarre experiences that occurred during my Vietnam combat tour always struck me as something out of a fantasy horror novel or movie. One reason for writing this memoir was to document these sinister events.

Also, over the years it became obvious that the Vietnam War had become ancient history to America's young men and women. They were most likely to learn about the war solely through popular culture, like the many inaccurate Hollywood war movies or television programs. A notorious example is the movie *Green Berets*, starring John Wayne. My readers can learn of the strange juxtaposition that occurred at our firebase while this unrealistic movie was being shown to the troops. Accordingly, another of my book's goals was to provide an accurate description of the war.

My Vietnam War memoir is a valid source of American military history based upon primary resources. The main resource was my memories of the war, which remained vivid because I frequently shared them with the combat veterans I worked with over the years as a clinical psychologist in the Veterans Administration's Health Care System. My memories were augmented by my private photographs and souvenirs, a Vietnam War slideshow I produced, and personal letters written to and audiotapes made for my wife. A quote from a July 1969 letter to my wife describes my attitude: "Now maybe you will realize that I don't like writing about the bad stuff. It is bad; I'd like to forget, yet that is what everyone is doing. Maybe I should remember, always. It has to change a person." The third goal of the book was to contribute to the collective written history of the Vietnam War.

My story is not a heroic portrayal of epic battles, but rather a tale of the difficult day-to-day slog through the muck and the horror of the war. These poignant stories describe my efforts to cope with the madness that was the 1969 Vietnam War in the Mekong Delta. For centuries wise men have known that fighting in a war as a soldier changes a man. For example, William Shakespeare wrote a portrait of the effects of war on the personality of one of his soldier characters, Hotspur, in his classic historical play *Henry IV, Part 1*. In my memoir I chronicle the transformation of a Roman Catholic altar boy into a brazen combat veteran, angry enough to challenge authority, afraid of nothing and everything at the same time. A young man's transformation from obedience to cynicism and skepticism led to clashes with the military's authoritarian culture, one not equipped to deal with college-educated enlisted men.

I want to acknowledge the contributions of my family editing team, consisting of my wife Judith, my daughter Colleen, and her husband Adam. This book would not have been written without their encouragement. I thank them for their assistance and advice from conception to completion of the project. I also thank my editor, Mr. Dylan Lightfoot of McFarland Publishing, for his guidance.

Walter F. McDermott, Ph.D.
Licensed Clinical Psychologist
Atlantic Beach, Florida
June 2018

1

Avoidable Casualties

We were flying swiftly into a dark night. More than just moonless, it was a primeval blackness, full of menacing evil. I sat on the hard metal floor of our single-rotor Huey helicopter, my feet dangling out of the open door. The wop-wop of the helicopter's rotor blade drowned out all other sounds. All that I could see below was the Mekong Delta's thick jungle, hurtling by like a dark rug being pulled from under me.

I was the lone passenger on a mission to resupply our battalion's artillery forward observer (FO). I knew that we were flying into a battle with the Viet Cong (VC). All that day I had been monitoring the battle over our military FM radios back in our battalion's tactical operations center (TOC) at our firebase near the village of Tan Tru, South Vietnam. One of our infantry maneuver companies from the 2nd of the 60th Battalion (2nd/60th) of the 9th Infantry Division had made contact with an estimated platoon-sized enemy unit. Our infantrymen had been chasing them all day, suffering causalities. Apparently the VC platoon had stopped running.

Immediately the colonel, our battalion's commanding officer, ordered a "donut operation," which meant he ordered his other three infantry companies to surround the enemy. They would encircle the VC and hold them in place by blocking all avenues of escape. Then we would destroy the VC in the middle of the donut by blasting them with artillery shellfire. The colonel landed his command and control (C&C) helicopter, manned by his headquarters staff, near the fighting, where they would be part of the blocking force and where he could manage the battle on the ground.

One vital member of the colonel's headquarters staff was his FO, an artillery liaison captain whose primary duty was to control the artillery fire into the center of the encirclement for the infantry. He would be in radio contact with our 105-millimeter and 155-millimeter artillery batteries back at our Tan Tru firebase and possibly with other artillery units within range. He was my immediate superior, my boss. Once the donut operation was ordered, he radioed back orders that I was to join him that night in the field to function as his radio telephone operator (RTO).

After recruiting another enlisted member of our artillery liaison team to monitor the radios at the battalion's TOC, I ran back to our bunker/barracks to pack my personal field gear and then to my captain's hooch for his field gear and finally to the supply building. It was all a rush job because the Huey resupply helicopter that would be flying me to the battle was scheduled to leave within the half hour.

The resupplies I carried were the critical extra radio batteries, which at that time were the size and weight of a large brick. Also I was resupplying our FO with extra C rations, canteens of water, two bandoliers of M-16 5.53-millimeter bullets loaded ten to a metal clip, and a deflated rubber mattress. Along with his resupplies I carried my fully automatic M-16 rifle, a PRC-25 FM field radio, two bandoliers full of clips of M-16 ammunition, C rations, and canteens of water. My job in the field was to carry the 20-pound PRC 25 radio on my back and to keep in constant contact with the firing artillery batteries and the Air Force's supporting fighter jets. It could be a dangerous job because the large FM antenna attached to the radios was a prime target for enemy snipers. I was relieved a bit when I saw that our artillery radios were equipped with the new and smaller folding antennas.

The battle was taking place at a remote corner of the Mekong Delta, in abandoned rice paddies near the Cambodian border. Only after boarding the Huey, armed with two M-60 machine guns, did I have the time to worry about what was about to occur. The darkness of the night produced a palpable sense of foreboding as I was flown to my destiny. My dread grew as we neared the battle site. My heart pounded madly, my stomach tightened in knots as the thought "hot LZ, hot LZ" roared through my brain. "Hot LZ" meant that the firefight, the infantry's ground battle, was ongoing and that our Huey would probably be the target of heavy enemy fire as I was being dropped off in the landing zone. I felt that I was being swept up by powerful waves of primal evil that I could not stop. I was helpless; there was no escape.

I attempted to steel myself for the approaching fight with the deadly Viet Cong. My breathing became rapid as I fought to keep my fear under control because I knew that panic in battle was fatal. But at the same time there was a type of anticipation going on in my mind. Not a positive type of anticipation, but rather a dreadful questioning of what exactly was going

1. Avoidable Casualties 5

to happen. This was a survival strategy that I had developed over my months in Vietnam. In combat I tried to think as fast I could, figuring out, as best I could, what exactly was happening and what would happen next. I was striving to outthink the enemy in order to increase my chances of surviving the war. As the anticipatory fear coursed through out my body, I asked myself, "How did a Catholic altar boy like me come to be in such a nightmarish situation as this?"

BAM, BAM, BAM. The door gunner on my right opened up with his M-60 machine gun on the Viet Cong enemy below. I physically recoiled from the unexpected firing so much that I almost fell out the helicopter door. With tracer rounds filling the air, the firefight was now visible. Red tracer bullets from our guns crossed in the air with the green tracers from the enemy's guns. Finally our Huey came to a hover about two meters above the middle of the muddy rice paddy. Most of the shooting was occurring near the right upper corner of the rice paddy about 15 yards from us.

In the Mekong Delta our Hueys never actually landed in battle because they were too vulnerable as targets when immobile on the ground. Instead, they hovered a couple of yards off of the turf, and then we soldiers were expected to jump off, no matter how high or heavily loaded we were. So I jumped into the battle.

With the weight of all my gear, I sank into the muck up to my knees. Since I was not an infantry soldier, I had not been issued a rucksack. I had to carry all my supplies in a medium-sized duffle bag I had somehow acquired. There I was, with ammunition bandoliers crisscrossing my chest, the PRC-25 radio on my back, my M-16 in one hand and a heavy duffle bag in the other, sunk up to my knees as the helicopter continued firing. The Viet Cong opened fire at my helicopter, trying to shoot it down. I tried to walk but fell, with the result that I was covered in mud. I was stuck and totally exposed in the middle of the rice paddy as bullets and tracer rounds filled the air around me. Even though it was night I recognized the faces of my FO and the other members of the headquarters unit frozen against the meter-high paddy dike wall. They stared at me. No one came out to help.

The sucking mud pulled at my boots as I crawled and stumbled the longest 15 yards of my life. Crouched over, my forward progress was slow; it was like walking in glue. As I struggled to the earthen mounds that defined the paddy's rectangular shape, the firefight was still going on. Our gunships were firing rockets directly in front of our rice paddy and making multiple machine gun runs.

Finally I reached the dike wall. As I hugged the dirt mound, I watched the infantry platoon at the dike wall perpendicular to ours on my right about 25 yards away, lying flat with their legs splayed out behind them. They were firing at the Viet Cong in front of them. I did not see any of the enemy, so I kept my head down and did not fire my M-16 automatic rifle. Just then one of our Huey gunships strafed the battle, blasting away with his machine guns from a high altitude down to a low level pass over the fighting. Tragically the gunner pulled his trigger a fraction of a second too soon, shooting the legs of one of our prone infantrymen as well as the enemy.

These types of appalling accidents were labeled in the military with the euphemistic term of "friendly fire," a term I learned to hate. Friendly fire was not friendly! My theory is that the label was designed by military lawyers to help camouflage the guilt of the men who shot the badly aimed fire. Yet friendly fire maimed and killed just as savagely as enemy fire; our wounded infantryman's pain was just as agonizing as if unfriendly enemy rounds had struck him. A more descriptive and more accurate term would be "misdirected fire." The Huey helicopter gunner had misdirected his machine gun fire; consequently he badly wounded one of our own infantrymen. He was guilty, but he would never be charged.

The unfortunate soldier's screams of pain penetrated the night air. The immense firepower of our helicopter gunships drove the Viet Cong back away from our rice paddy perimeter. The word came up the line that the casualty was a Black soldier. He screamed and screamed as one of our medics made his way to him and administered basic medical care. As I heard the injured soldier crying out again and again for his mother, a cold chill ran down my spine. His screams made me realize an unsettling fact about his maturity and that of most of my American comrades in arms: "These are boys, not men but BOYS!" Years later I learned that the average age of an American combatant in the Vietnam War was 19. Compare this age to the average of World War II American soldiers, which was 26. I was hearing an adolescent crying out for his mother to come and take all the pain.

My age at the time was not that much older, only 23. But I had already paid my own way through college by working summers in a factory, had graduated summa cum laude, and was engaged to marry my college sweetheart. I had lived on my own. I was a man, a young man certainly, but not a boy who lived with his family. I was not a boy who had his mother cooking for him and cleaning his dirty laundry. I came to the harrowing recognition that my surviving this grotesque war depended upon such inexperienced and emotionally vulnerable boys.

1. Avoidable Casualties

As the fighting died down in our immediate area, the medic was able to radio in a request for a "dust-off," which was a nickname given to medical evacuation Hueys. Before long the dust-off helicopter arrived on station. Its pilot then performed a feat of raw courage as he actually landed his aircraft in the next rice paddy to our right. The badly wounded adolescent soldier was carried to the dust-off by his comrades and successfully evacuated. Now, because of the relative calm, I was finally able to connect with my FO captain to resupply him.

After that we all hunkered down in place and tried to get some sleep right there in the mud of the abandoned rice paddy. Even though I had delivered to the captain the air mattress he slept on, I had not been issued one. Instead I had to try to sleep on the mud with my upper body propped against the paddy dike wall and my legs sticking straight out. Exhausted from the physical and psychological stress, I feel asleep quickly, but it was not a restful sleep. I was repeatedly awakened as the mud sucked my legs down into the muck. This irritating pattern of falling asleep and then sinking into the mud repeated itself during most of the night. To fight the mud's grip I pulled myself higher and higher onto the paddy dike mound. Finally, because of fatigue combined with sleep-deprived muddled thinking, I threw myself over the dike and slept with my stomach draped over the wall.

"McDermott! Are you all right? McDermott! Are you all right?"

Opening my eyes in response to someone screaming my name, I saw that it was my FO shouting at me. He had awakened on his air mattress and seen my body slung over the dike. To him I looked like a corpse. He was afraid that I had been killed during the night. When I reassured him that I was stiff but alive, he was relieved but irritated.

By the morning the fighting had moved on as the Viet Cong withdrew. During the night most of the enemy platoon had escaped our encirclement by utilizing one of their standard escape strategies. The Viet Cong had attacked in force the two small infantry outposts that bordered a good-sized canal and had overwhelmed the defenders. While these outpost firefights raged on, the enemy floated the rest of their unit and weapons down the middle of the canal. My angry thoughts at the time were that the enemy's escape plan should have been anticipated and countered. Officers would argue that there were simply too many canals and waterways in our Mekong Delta battle area to reinforce every outpost that bordered on one. The infantrymen who had to actually defend those vulnerable outposts would scorn those arguments.

It was another oppressively hot day in the Mekong Delta, filled with intense heat, sweltering humidity, and nothing to protect us from the blazing sun. Sweating through the morning, we ate a breakfast of cold C rations. The commanding colonel ordered our artillery headquarters platoon to return back to our firebase near Tan Tru village. Charlie Company, who had fought for our rice paddy last night, was given the dangerous assignment of pursuing the fleeing Viet Cong enemy unit. They dutifully formed up, loaded their weapons and trudged out with slow deep steps into the forbidding jungle that the Viet Cong had fled into. As I watched them disappear into the thick bush, I admired their courage, and yet I was grateful that I was not an infantryman, the most dangerous military occupation.

As we gathered our gear and waited for the helicopters that would be flying us back to our firebase, we heard over the radios that a sister infantry unit, Delta Company, had discovered a large cache of Russian rockets. In a war devoid of advancing front lines or conquered territories, our upper-echelon officers viewed these enemy weapons caches as major trophies. Instead of dealing with the discovered weapons in the safest manner, which would be blowing them up in place, a colonel had ordered that the ordnance was to be delivered to his headquarters by helicopter. He probably planned to have the enemy weapons sorted and piled neatly at his base camp in order that he could have publicity photographs taken as he proudly stood over the pile. This type of asinine order was the result of our upper-echelon officers' desperation to demonstrate to their political superiors some type of progress toward victory in the Vietnam War. The only war trophies they lusted after more than weapons caches were large stacks of dead Viet Cong or North Vietnamese Army bodies.

A short time later the news came over our radios that one of Delta Company's soldiers had dropped one of the rockets he was carrying. The rocket had exploded, blowing off both of his legs and a leg of the soldier carrying rockets behind him. There was blood everywhere; the legless soldier died within five minutes. The other poor guy was flown to an army hospital. Both soldiers had been dutifully carrying out that colonel's ludicrous orders.

The news angered and depressed all the soldiers in our unit; we were all upset over this needless death and grievous injury. Our Huey helicopters finally arrived and flew us back to our firebase. I spent the entire flight grappling with the absurdity of the tragedy. Why did it happen when it could so easily have been avoided? Even though I did not know the poor man who lost his leg, I felt sorry for him. He did not even have the small consolation of being hurt while courageously fighting the enemy. Even if he survived, what would his life be like? Would he be confined to a wheelchair for the rest of his life? How could he come to grips with such a dreadful loss?

After landing and getting settled at our firebase I relived the entire night in my mind. My conclusion was, "Another crappy no-win operation plus preventable deaths," triggering a shiver of despair.

2

Basic Training

In 1968 America's Vietnam War reached its ugly nadir. It was the year of the infamous Tet Offensive and the year that produced the highest casualty numbers among American servicemen fighting there. It was also the year that I graduated from college. After working my way through college by toiling as a laborer in a Pittsburgh factory, I was able to graduate early with highest honors from Indiana University of Pennsylvania (IUP). My major was psychology and my plan was to study clinical psychology in graduate school. But the federal government had a different plan for me. During the Vietnam War the United States still used compulsory conscription, better known as the draft, to fill the ranks of its armed forces.

The federal government intended to reach its draft quotas of able-bodied young men who would be forced to serve in the military and possibly sent to fight in Vietnam. Young men today do not understand the anxiety that the young men of my generation experienced because of the draft, this Sword of Damocles hanging over our heads. If a young man refused to be conscripted into the military, he could be sent to prison. Yet in big cities like Pittsburgh, young men who had a connection to a political figure like a city councilman or a relative on the draft board could be assigned one of the coveted openings in a local military reserve unit. During the Vietnam War 99 percent of reserve units were not called up to fight. As one of the poor masses without political connections, I was envious of those lucky men. Other desperate young men faked psychiatric or medical conditions in attempts to fail their draft physicals. Many cowards fled to Canada, which offered a safe haven for draft dodgers. Men in college were given student deferments as long as their grades were acceptable. But there were mounting criticisms against all such student deferments, which were said to favor the wealthy. This argument did not apply to the male students at IUP, who were, like myself, children of blue-collar workers. To partially

2. Basic Training

mollify these critics, federal officials cancelled all graduate school student deferments in 1968.

After graduating in January of that year, I was fortunate enough to secure a job teaching psychology at the high school I had graduated from, Shaler High School. During those few months of teaching, I was accepted into three well-respected clinical psychology graduate programs. Consequently I requested that my local draft board grant me a postponement of my draftee status. They only agreed to a one-semester postponement. I was crushed. I knew that there was no way I could cope with the demands of graduate school with the threat of the Vietnam War hanging over my head.

Their decision forced me to consider my military options. Since I was a college graduate, I was eligible for Officer's Candidate School (OCS). If I had to go into the military, being an officer seemed like the best option. At that point I thought that serving as an officer would be a favorable addition to my résumé. After meeting with the local Air Force recruiter, I applied for their OCS in navigation because I had always been proficient in land navigation using a compass and maps when I was a Boy Scout. Unfortunately, I failed the Air Force's physical examination because of my flat feet. However, I had not anticipated that my Air Force physical would become my draft physical and consequently move me up to the front of the draft line. I was now classified as 1A, that is, subject to immediate call-up. I became desperate.

As a secondary option I applied for Army OCS in artillery, as recommended by my Army recruiter. Given how much soldiers march, it was ironic that the Army accepted me even with flat feet. The recruiter promised that I would be issued a special type of boot to compensate. That summer I received my first military orders to fly, via commercial jets, to Fort Leonard Wood, Missouri for eight weeks of basic training, a requirement for all soldiers, including officer candidates. For the occasion I dressed formally, wearing a white shirt and tie along with a blazer because it was the first time in my life that I had ever flown on an airplane.

My miserable military experience started on June 28, 1968, at "Little Korea," the nickname of Fort Leonard Wood because of its extremes of weather: searing heat in the summer and frigid cold in the winter. The extra harassment I would receive in basic as a college graduate started within the first hour. The sergeant who collected me and the other men in my situation harangued us as fools for not joining our colleges' Reserve Officer Training Course (ROTC). Instead we all had to endure eight weeks of basic training, then eight more weeks of advanced individual training

(AIT), and finally six months of the rigorous officer's candidate school. ROTC training was significantly less demanding. But what could we do? We could not turn back time. We were all stuck on our present path. Thus began the unending degradation that continued throughout my brief military career.

Before actually starting basic training we all had to suffer the indignities of a weeklong transition into military life. During this transition time we were first sworn in to defend the Constitution; then we had our heads shaved and were issued uniforms. One particularly stressful requirement was that of receiving multiple injections of vaccines against various diseases. We all had to roll up our sleeves and march between lines of medics, who simultaneously shot the vaccines into both of our shoulders with pneumatic air hypodermic needle guns. One had to be perfectly still, because if one flinched even a little under the air gun injection, it would cut open the shoulder muscle. I saw men cry out in pain as their shoulders were ripped open by these injections. A few men fainted. Each of us received ten different vaccines in this manner.

By far the most stressful aspect of the pre-training week was life in the barracks. All new trainees, both OCS candidates and draftees, were housed in old wooden barracks. We were all just trainee maggots to the Army trainers. Among the draftees, the loudest and wildest were a large group of Blacks from Chicago, many of whom were apparently members of infamous street gangs like the Blackstone Rangers. This particular street gang had received national media coverage after being caught stockpiling grenades and a bazooka. One tall, heavily muscled Blackstone Ranger shouted out to an entire brigade formation that after seeing all the weapons the Army possessed, he had to admit that the Army could beat the Blackstone Rangers in an all-out street fight. These gang members dominated the barracks at night, running around yelling and bullying other trainees as if they were on their first gang campout. They deprived all of us of much needed sleep, yet no one dared confront them. At night I heard other recruits sobbing in their bunks.

Worse yet were the bathroom facilities. There was no privacy for a recruit moving his bowels. The twelve toilets were open and arranged in a semi-circle. We also all showered together, which was no problem for me because we had group showers in high school football. But we always had private toilet stalls. At first I tried to avoid the toilets, but eventually the call of nature could not be ignored any longer. It was terribly embarrassing, trying to defecate while facing ten strangers sitting on nearby toilets doing the same, accompanied by all the smells and sounds of such an endeavor.

2. Basic Training

There was no joking, little if any talk at all. I kept my mouth shut and focused instead on finishing my business quickly. It never became any easier.

At last the eight weeks of basic training officially began. We were all marched to another section of the fort and assigned to a new training company unit. I was assigned to a training company composed mainly of men who were fortunate enough to have joined Army reserve units in their small southern towns. Their enlistment contract required only 16 weeks of basic and advanced individual training; they could then return to serve the rest of their enlistment in their home towns, thus avoiding Vietnam War duty. I was resentful and jealous of these men. In Pittsburgh, my home town, a man's family had to have political connections in the Democratic machine to be assigned a coveted spot in the reserves. The remainder of the company was made up of a few of us OCS candidates. We were all housed in a metal barracks that looked like a huge corrugated steel pipe that had been cut in half. All the bunk beds were set up in one open bay; in addition there was one separate room where our drill sergeant lived, as well as a bathroom of sinks and toilets. It pleased me that the toilets were in individual stalls arranged in a straight line, even though they had no doors.

Every day of basic training was long and physically taxing. Before dawn the drill instructor (DI) blew a whistle and banged a metal garbage can lid on bunks to startle us awake. We hurriedly donned our uniforms to assemble outside for a three-mile run. During the first week an overweight trainee twisted his knee on the run; yet the drill instructors forced the crying man to continue, damaging his knee even worse. One morning I saw trainees in another company ordered to literally drag a trainee by his feet on their run. Still sweating from the run, we had to move quickly through the cafeteria line for breakfast. We were required to wolf down our meal in the noisy and crowded mess hall. No socialization allowed.

The rest of the day was filled with some type of military training. There were days filled with throwing live grenades and others spent learning how to shoot our M-14 rifles. One nasty day we were forced into a structure where poison gases were released, while we removed our gas masks and shouted out army slogans before being allowed to don them again. I can still recall how deadly chlorine gas smelled like a swimming pool. My exposure to tear gas made it difficult for me to open my teary eyes afterwards. Most of the time we practiced marching in close order drill carrying our rifles, while following the orders and cadence the DI shouted out. We spent hours on dusty fields in the hot and humid August sun, sweating and marching, sweating and marching. Also all trainees had to work one day of kitchen police (KP) duty. On my day I reported to the mess hall sergeant in the dark hour before regular reveille. He assigned me to the worst job, back sink, which meant cleaning large grease-stained pots and pans all day long and into the night. It was hot, fatiguing work for a 16-hour shift. I did not return to my bunk until ten o'clock that evening.

Another stressful factor that was difficult for me to cope with was the sleep deprivation, which took both a physical and a psychological toll. I was always sleepy during basic training. First of all I was not used to the hours, going to bed at nine in the evening and then rudely awakened at four in the morning. Besides there always seemed to be some extra duty assignment that interfered with the time allotted for sleep. For example, Army regulations required that someone always be on guard in the barracks at night for fires. The fireguard shift was only two hours a night; thus four basic trainees had to share a night's duty. The four-man requirement meant that I was assigned to fireguard about every other week, which meant being awakened at odd hours. There were days when our company was assigned night guard duty for the entire fort, which resulted in four hours of sleep that night. At other odd times the DI would awaken us early as punishment for not performing perfectly in that day's training or for not cleaning the barracks well enough. One morning we were awakened early and ordered to stand in formation outside our barracks. We all had to stand at attention for half an hour out in the warm, pleasant morning air. Within a few minutes I involuntarily fell asleep on my feet. I only awoke when other trainees bumped into me after the battalion had been ordered to march. I had thought humans could not sleep standing up, but on that morning I did.

Our DI was a young southern white Sergeant First Class (E-6) who had fought in Vietnam the year before. He loved to frighten us with his war stories, one of which truly horrified me. The sergeant reported that at times the VC communists would ingest meta-amphetamines before executing a ground attack. They would also tie bands on their arms and legs that, if they were wounded, could be tightened into tourniquets. These chemically stimulated guerrilla fighters would then charge into an American defense perimeter. Even if an American soldier scored a lethal hit on one of these VC attackers, because of his amphetamine ingestion the enemy could still run a few more steps, pulling on the trigger of his AK-47, which continued to fire. Those bullets could kill an American soldier just as dead as those fired by a living enemy. Being shot and killed by a dead man; the concept was so eerie that it unnerved me.

I tried to keep a low profile during basic training, but it was difficult.

2. Basic Training

I could not hide my contempt for the entire process; my disgust was visible. Early on, our training company's commanding officer singled me out for extra harassment. As a young man I always had a heavy beard. It became an ongoing conflict between the commander and me. He repeatedly accused me of not shaving that morning, when in fact I had. He would order me to shave right then in front of him while he yelled at me the entire time. One of his frequent harangues was that I was a "hippie college student"—far from the truth—and that I had only joined the Army to write a book about it. At that time I had absolutely no intention of writing a book; it is ironic to be actually writing a memoir now, close to 50 years later.

On August 11, our training battalion commander ordered us to hike 20 miles into the Fort Leonard Wood's thick forest areas. We bivouacked in two-man pup tents fashioned by joining two poncho liners together. These pup tents were so small that my tall tentmate's feet stuck out from the tent when he slept. It was rough trying to fall asleep on the ground with only a tarp covering the hard dirt. The training commander accused me again of not shaving the next morning, when in fact I had. He became so angry that he ordered me to stand at attention before my entire company while he dry shaved me, that is he scraped my dry face—no soap or gel was applied—with a dry razor. It was both a frightening and a humiliating experience, especially when he yelled out, "Here is a college graduate who cannot shave himself." The concept of a college graduate enlisted man angered this officer. My current theory is that my degree challenged his personal sense of adequacy; perhaps he had never graduated from college. At that point I tried to remain perfectly still, even though I was fearful of this enraged man scraping a razor across my face. Besides burning my whole face with that dry razor, he made a few small bleeding cuts into my face before he summarily dismissed me.

One night in the bivouac some other company's DI ordered a group of us trainees to run in our leather Army boots. During the long hike to our bivouac, my boots had rubbed my skin into a large blister on my right foot. On that night's run the blister broke open and my raw skin started bleeding from my boot, irritating the wound. The pain forced me to drop back in the run until finally I fell and rolled into the bushes lining the road. I tried to lie still in the bushes to hide, but within a few minutes the enraged DI found me. He ordered me to run back to our campsite with him, an intensely painful run because of my bleeding blister. Once we reached camp the DI demanded to know why I had dropped out of the run, so I explained my foot injury. He then ordered me to take off my right boot. After I removed the boot, I turned it upside down to drain my blood. The amount of blood pouring out shocked the DI into realizing that I had a legitimate complaint. He ordered me to report to medics the next morning for treatment of my foot wound.

On the long march to the bivouac site we had to carry our provisions and survival gear in military backpacks. Since we were forbidden to bring extra food, my small resistance involved carrying three cans of fruit cocktail in my backpack as my own secret treat. I felt that I had to resist the continued oppression in order to preserve my identity. Once darkness fell on that first night, DI supervision lessened. I invited a few of my trainee friends to my tent to share the forbidden fruit cocktail. My pup tent party was a big success. We all enjoyed the treat and I reveled in my rebellion, explaining to my friends, "We are not animals, even though they treat us as animals." The next night our camp was "attacked" by an enemy unit composed of DIs and some other training cadre. Dozens of my fellow trainees took the staged attack seriously, getting into the spirit of the fight by shooting noisy blanks at the attackers. I could not be bothered with such games; instead I stayed put in my pup tent trying to sleep as long as possible.

The constant harassment and stress of basic training wore me down psychologically. The Army nearly crushed my spirit. In desperation I tried to cope. My first response was to call upon my religion. I was raised in a strongly religious family who prayed the Rosary together out loud every evening along with a Rosary prayer radio program. I was both an altar boy and a choirboy for our parish church. So in basic I prayed every morning and evening. I prayed whenever I could, even on marches. I would recite the Rosary over and over again. I was obsessed with prayer; during every break in the training I would pray. I prayed and prayed for the special miracle of a discharge from the Army.

The stark prospect of going to Vietnam to kill upset me. I thought that killing people conflicted with my religious beliefs. Was not one of the Ten Commandments "Thou shall not kill?" When I discovered that even a recruit could apply for conscientious objector (CO) status, my desperation was so severe that I applied. Of course my application upset my training cadre, turning them against me even more. I was ordered to meet with an Army chaplain for an interview in his office. I was intensely anxious and tense the entire interview; to me the chaplain was neither a priest nor a minister, but just another Army officer in full uniform. The interview did not go well. Finally he asked why I was so afraid of him. I was so nervous that I could not articulate a valid answer.

Next I was ordered to see an Army psychiatrist for an evaluation.

2. Basic Training

However, since at this point it was already near the end of my eight weeks of basic training, my company commander decided to postpone the psychiatric evaluation until my next assignment, which would be at another base for advanced individual training. My commander placed all my conscientious objector paperwork in a large manila envelope, while happily informing me that he had included a detailed description of all the weapons training I had received in an effort to sabotage my application. He then ordered me to hand carry the paperwork to my AIT assignment.

During my first week of basic training, I had attended the Sunday religious service. It was not the standard Roman Catholic Mass that I expected. It turned out to be a non-denominational generic Christian service run by a chaplain. The service was foreign to me, like some type of bad amateur hour show with all the singing. The next week I learned that some recruits had found a post exchange (PX) store on the base that served beer on Sunday mornings. Since it was one of the few times that we were not dominated by our DI and there was not a real Mass to attend, I took advantage of the Sunday lull and joined my buddies at this PX. There were picnic tables outside the store where we sat in peace and enjoyed our drinks. I was there, not because I loved beer, but rather for the opportunity to relax and socialize with my fellow recruits. I never again attended another religious service during my stateside training.

The pressure of basic training continued unabated. One afternoon, after a long day of training, our company returned to our barracks only to find that training sergeants had trashed our barracks, knocking over all the bunk beds and lockers. They had taken our personal footlockers and tossed them into a big pile in the middle of our common room. All our boots and uniforms were piled together in a huge heap. It was so bad that even our DI became angry. As tired as we were, we still had to sort out and clean up the mess. It took hours to clean it all up, again limiting our sleep time. Another time I was given extra punishments because I awoke in the middle of the night to write a letter to my college sweetheart under a flashlight. In my letters home I reported on a week in basic when I had at the most five hours of sleep a night.

I was always searching for peace and quiet. One oasis was the base library, where I discovered that no one would bother me while I was reading there. The library was seldom used; consequently, the soldiers in charge were glad to see anyone, even a trainee, show up and take advantage of the books. After this exciting discovery, I routinely skipped the mess hall dinner, bought a candy bar at the PX, and escaped to the library for peace. It had a collection of the Great Books that appeared to be brand new. I found the works of Sigmund Freud in that collection. I would relax and read about Freud's psychoanalytic theories while contentedly munching on my Baby Ruth candy bar.

As the weeks of basic training continued, the stress increased. My prayers were not answered. There was no miracle and I was desperate for a way out. I had to find a solution. My thinking was so distorted at that point that I started to read the Encyclopædia Britannica's articles on human biology in order to find a method of injuring myself that would lead to a medical discharge. Shooting myself at the rifle range was one option considered, but I concluded that it would be too difficult to limit the ultimate physical damage from such an assault or deal with the inevitable pain. After considering a number of other injuries, I settled upon rupturing my eardrum. I read everything I could find on the human ear, attempting to find the perfect way to rupture my eardrum so that it would appear accidental and at the same time minimize the actual physical damage. It took me weeks to formulate the beginnings of an elaborate plan. Planning brought me some comfort; I was not as helpless as I had been feeling, and it could be a way out of the Army. I needed this fantasy so that I would not totally lose my mind. I never acted out the fantasy.

Towards the end of basic training I began to seriously question my decision to attend OCS. By that time I realized that a commission as a lieutenant was not the exalted position I had once believed. A lieutenant was in fact the private of the officer. Now that I was part of the military, I saw how lieutenants were assigned so many duties and responsibilities that it was difficult for any one man to do all of them adequately. They were assigned the duties that the higher ranking officers rejected, like being in charge of a mess hall. His many stateside duties insured that the lieutenant had little time to sleep. Plus I learned that the United States Congress awarded an officer a commission for life. If needed, Congress could order an officer back into the military. For example, during the early 1950s many World War II officers, who were just starting to achieve some success in their civilian careers, were called back into the military to fight in the Korean War. That prospect repulsed me; once discharged I wanted to be finished with the military, period.

If I were to quit OCS, my enlistment contract required only that I finish my two-year enlistment. Also, if my active duty included Vietnam War service, then I would be exempt from any active reserve duty. Basically, after a tour in Vietnam, my military obligation would be fulfilled. If I were to remain an enlisted man, there was also the added incentive of a safe military occupation. At our Sunday morning beer socials, the other OCS

2. Basic Training

candidates informed me that our AIT training would be in fire direction and control (FDC). My friends explained that an FDC specialist utilized trigonometry to aim the big artillery guns. It was considered a relatively safe combat assignment. Usually the FDC specialists worked in a fortified bunker near the artillery howitzers. They were not required to go out on patrols outside of their fortified positions. Infantry companies provided security for the artillerymen.

In contrast, a new second lieutenant artillery officer usually functioned as a forward observer (FO) for an artillery battery. This was a dangerous job in the Vietnam War. The FO officer was required to go out on combat patrols with infantry companies in order to call in targets back to his artillery batteries. The FO's life was as much at risk as the life of any infantryman.

Finally, there was the extra incentive of the Army's early-out program. If a soldier was accepted to a college or university, he could be discharged five months early to start his higher education. Since I already knew that I could enroll at Indiana University post discharge, I was eligible for that five-month drop in my enlistment. All these various considerations led me to my firm decision that I did not intend to suffer through 26 weeks of OCS harassment for some nebulous military prestige. My decision was to quit OCS as soon as possible and try for the college early out. I did not share my decision with anyone.

Over those eight long weeks of basic training my discontent and anger grew and grew. Toward the end I was ready to explode. One time at the firing range we were issued live rounds for our M-14 battle rifles. With the rounds in my weapon, I contemplated how to turn around and shoot the company commander. I only had a few bullets, and he was savvy enough to stay far back from the firing line. I never had a clear shot and had to dismiss the idea, no matter how appealing.

In the last week of basic our company spent days cleaning and polishing our barracks for our final inspection. The DI had appointed some of our company's trainees as acting sergeants, supposedly to teach them leadership skills. These temporary promotions went to the heads of a few of the men. One such jerk was a six foot four inch tall, red-haired reservist from Tennessee, one of those men we regular army recruits were jealous of. I was already mopping the barracks' floor when he approached me. He ordered me to cut the grass outside our barracks with scissors and attempted to hand me a pair. If he had given me a lawn mower, I would have obeyed his orders; but the use of scissors demonstrated that it was all about humiliation, not about completing the work. It also troubled me that this stupid order came from just another recruit, not a real sergeant. My anger boiled over. Even though he was taller and heavier, I balled my fists and told him that if he handed me those scissors, "I am going to shove them up your ass." I meant it. He recognized that I was seriously ready to explode into violence. He backed down, turned and walked away quietly.

My sense of helplessness lessened as basic training wound down. This hell of organized torture was going to end soon. Yet my anger at the petty tyrants in the higher ranks remained. It was a constant struggle for me to control my hatred of our training cadre as well as the entire authoritarian military system. My outrage peaked during our company's practices for graduation from basic training. The Army's use of the term "graduation" rankled me. After years of intense study, I had graduated from a university; that was a true graduation, not this parody. But still the top sergeant of the training battalion made us practice marching in the hot summer sun on a dusty dirt field. We marched and marched in the heat and humidity until there was a cloud of dust hovering over and on us.

The big deal was that there would be a general attending the ceremony. We were required to march in review past him. The top sergeant, who was as big as any professional football player, was arrogant enough to act as the general's stand-in. He stood at the podium in front of the battalion's formation. All the companies were to march past him in order to salute him with an "eyes-right." This command meant that every man in the company, except, for some unknown reason, the row closest to the general, would snap his head to the right to face the general while marching past him. When it came for our company's turn, I marched with my fellow trainees to the front of the formation and then our drill sergeant gave the "eyes-right" command.

In clinical psychology it is called an "irresistible impulse." I did not plan my reaction to the order; nor did I consider consequences or penalties. An impulsive idea suddenly came to mind; all my years of training in Catholic inhibitions could not prevent me from acting out this impulse. I snapped my head to the right and stuck out my tongue at the top sergeant pretending to be a general. His reaction was fascinating to watch. If I had punched or kicked him, he would have had ready-made responses programmed into his primitive brain. But with my affront he had no programmed response. I saw his confused face as he decided what to do. In the meantime I kept on marching as if nothing had happened, but to myself I kept whispering, "You are going to get it … you are going to get it." Our company marched back to its original position in the battalion formation. I was standing at attention in the formation when suddenly I found myself flying through the air. The top sergeant had snuck up from behind, grabbed my collar and my belt, and flung me into the air. I flew about ten yards before crashing to the ground. He yelled, "You are not graduating. Go report for KP."

I picked myself up, dusted off some of the dirt and walked away from the formation. By that time I knew that the top sergeant had no means of communicating my banishment to the mess hall sergeant. So instead of reporting for KP, I went to the PX to buy a couple of candy bars and a Coke. Then I returned near the field used for graduation practice and hid under a tree. I spent the rest of the afternoon lying pleasantly in the shade, sipping my Coke and munching on my candy bars while watching my fellow trainees marching around in the baking sun and clouds of dirt. It was one of the high points of my military career.

I never formally graduated from basic training. Because I was sure that the top sergeant had communicated with the mess hall sergeants by then, the next morning I reported for KP duty at the mess hall before dawn. Interestingly, the mess hall sergeant assigned me to the easiest of jobs, wiping clean the officers' table after they had finished eating. I cannot help but wonder if the mess hall sergeant found my transgression amusing. For most of the day I just stood around in air-conditioned comfort. That night I discovered that, after the graduation ceremony, all the trainees had had to clean their rifles and military equipment to pass a rigorous final inspection. Since I was on KP, other recruits were ordered to clean my rifle and pack up my equipment and belongings for me.

On the next and last day of basic training each trainee was given his new orders, which were read out loud by the top sergeant. Most of the draftees were ordered to infantry AIT. When I was given my orders to report to artillery AIT at Fort Sill, Oklahoma, the top sergeant laughed out loud. He understood that my orders meant that I had a combat military occupation specialty (MOS), which meant that I would eventually be ordered into the Vietnam War. This was late 1968, after the communists' major Tet Offensive; it was not a surprise that I was on my way to the Vietnam War, yet I resented the top sergeant's laughing about it. It was no joke.

Once my basic training was completed, I finally had time to decompress on my one-week leave. I flew back to Pittsburgh—the second time in my life that I had ever flown on an airplane—to stay with my family and spend time with my college sweetheart. Yet when we went out to dance I felt like a weirdo because of my very short military haircut. My only saving grace was the fact that I am a big guy and people thought that I was a football player in training. The week flew by far too quickly.

3

Advanced Individual Training

My new orders were to report for duty at Fort Sill, Oklahoma, the U.S. Army's primary artillery training base. Every soldier is given specialized training during his advanced individual training in his assigned military occupation specialty (MOS). My MOS was in fire direction and control (FDC), which is the job of aiming the artillery big guns through the application of trigonometry. Its official military code is 13E20; for comparison, the infantryman's MOS code is 11B, which was often used in the Army as a synonym for infantry.

Once I reported for duty, I soon realized that my AIT experience would be different from my basic training one. First of all, we trained in platoon-sized units—roughly 20 men—as opposed to the company-sized units—roughly one hundred men—in basic training. Generally in the U.S. Army, the smaller the group, the more specialized and elite the unit. Secondly, all the men in my training brigade were college graduates on their way to artillery OCS. Within this group of trainees I felt more comfortable because their attitudes were similar to my own. Another major difference was that our drill sergeant was rather humane. He was a large Samoan from Hawaii who possessed a relaxed attitude towards his trainees. As long as we kept our barracks clean and squared away (the Army's term for "neat") and passed all our required tests, he would avoid all extra harassment. At night he drove a taxicab in nearby Lawton, Oklahoma to supplement his meager Army pay.

However, the most striking difference in AIT was the Army's emphasis on learning the necessary FDC skills. Most of our time was devoted to classroom training. To plot our firing problems, we utilized large drafting tables and T-squares like those used in mechanical drawing. The artillery's

3. Advanced Individual Training

big guns are officially termed howitzers because they are specifically designed for indirect fire, which is firing at a target that the gunner cannot see. A howitzer's design contrasts with that of a cannon, which is designed to fire at a target that the gunner can see. The advantage of a howitzer is that its barrel can be raised to a higher angle than that of any cannon, thus allowing the howitzer to fire over obstacles like a hill or a building.

The basic concept for aiming artillery howitzers is that the FO, the howitzers and the target form the corner points of an imaginary triangle. The howitzers in their firing positions are at a known location. If he can accurately read a map, the FO's reported position would also be known. With this information an FDC can calculate the distance between the guns and the FO. The FO can see the target directly, so he can estimate the distance between him and the target, or he can pinpoint the target's location through grid coordinates on his area map. From this information the FDC can calculate the distance from the FO to the target. Thus we have described two sides of the triangle: the line between the guns and the FO and the line between the FO and the target.

Next the angle formed by these two sides of the firing problem triangle can be calculated. Trigonometry provides the mathematical functions that enable the FDC specialist to use these two sides and one angle of our imaginary triangle to calculate the range to the target and the required angle for the gun's barrel to ensure that the howitzer's rounds hit it. This is a simplified version of the explanation. There are other considerations, such as the weather conditions and/or differences in elevation, that could be included in the FDC's calculations. The trigonometry calculations were performed with the aid of large specialized slide rules, which were the state of the art analogue computers of their day. Remember, this is in 1968, a time when pocket calculators were only a fantasy in the minds of mathematicians.

During the last weeks of our training, our instructors introduced us to the Army's new desktop FDC calculator. It was as large as an old cash register and was only capable of one mathematical function: artillery firing solutions. Yet it was the state of the art for digital calculations at that time. We played with the calculator in the classroom, but did not use it on any of our live fire practices. In Vietnam I talked with FDC specialists who had used the same model of desktop artillery calculator. They reported that the battery commander did not trust the accuracy of these new digital calculators. He ordered his FDC crew to double check the machine's results using the old but trusted slide rules. Ninety-nine percent of the time the double check upheld the machine's accuracy. However, when the results were not duplicated, the commander insisted upon a triple check of all the calculations. My friends reported that inevitably the triple check indicated that the machine calculator's results were the correct ones. Today all these artillery firing problem calculations are performed on the latest military digital laptop computers, aided by the precision of global positioning satellite information.

Later in our training we had actual live fire exercises. To make it more stressful on us, our instructors set up our FDC position between the guns and our targets, which were old tanks and trucks located on a hillside to our front. If we were to make a serious error in our calculations, there was a chance that the live artillery rounds could strike us. This arrangement certainly motivated all of us trainees to double check our results to ensure their accuracy. Our instructors role-played the FO function of calling in target locations. We calculated the ranges and firing angles and then radioed them to the men manning the howitzers; we called these soldiers the "gun bunnies." It was exciting, calling in our solutions, hearing the guns firing and actually seeing the rounds hit near their designated targets. We would then act as the FO, calling adjustments to the guns to ensure that the next rounds struck squarely on top of the rusted tanks. When the first rounds hit squarely on a target, we all celebrated. Accurately controlling these powerful howitzer weapons made us feel like giants. These live-fire exercises truly forced us to learn our craft well.

Our FDC training platoon was regularly tested in the classroom. Many of the tests were elementary problems in algebra. Some men actually studied the official training manuals for the tests, but I never bothered. By paying attention to the instructions in class, I learned all the basics I needed to make an intelligent effort on the other type of tests, which were actual artillery field firing problems. Thankfully I did well on all the tests; in fact, our platoon set new brigade records for the best test scores. Our superior performances in the classroom kept our drill sergeant happy, so he kept extra harassment to a minimum.

But there always was some harassment. In a pitiful attempt to teach leadership skills, other FDC trainees, who were only a few weeks ahead of us in training, were given the temporary rank of sergeant. The real sergeants would order our platoon into after-dinner formations and then have the pseudo-sergeants ask us questions about various army regulations and artillery specifics. If any trainee failed to provide the correct answer, the temporary sergeants would punish him by ordering him to do push-ups. The ritual was to question us while we were braced at attention, standing tall and straight. If one of us failed a question, the temporary sergeant would yell out the command, "Hit it," which meant that the trainee was to immediately fall straight to the ground into the push-up position and then perform his punishment push-ups.

These after-dinner formations became routine and generally were not horrible. Unfortunately, some of the pseudo-sergeants let the temporary rank go to their heads, turning them into Napoleon-like tyrants. One such Napoleonic jerk was harassing me with picayune questions. He could not even look me in the eye because he was so short that his head only reached up to my chest. After I failed one of his questions, he yelled out the order, "Hit it." Like a good soldier I immediately fell straight forward. This undersized tyrant's reaction times were abnormally slow, because after yelling, he made no attempt to move out of my way. I fell directly on top of him, knocking him to the ground. The supervising drill sergeant went ballistic with anger, screaming, "I can't believe you did that!" He ordered me to do many extra push-ups, but he was limited in his punishments because I did follow the command exactly as required. When I finally returned to our barracks that night, I received a loud round of applause from my platoon mates. I was considered a hero by them, even though I had not made a conscious decision to fall on the tyrant. I was simply following orders.

I did receive some further mild punishments over the next few days. After long classes our platoon was usually given a few minutes of break time. During these breaks, I was not allowed to relax. Instead I was ordered to stand at attention at the front of the classroom. It was all worth it because of the high status I now possessed with the rest of my platoon. Most of them resented the harassments from the pseudo-sergeants; they admired my resistance.

There were other instances of extra harassment in AIT. One day our platoon returned to our barracks from a field exercise with our DI, only to discover that someone had trashed our barracks. Just like what had happened in basic training, our bunks and lockers were all toppled over with their contents spilled out everywhere. We later found out that drill sergeants from other platoons had done the dirty deed. Our good-hearted DI had not even been forewarned about the attack. He became angry over the trashing, but we never learned how he responded to the misdeeds of the other sergeants. Regardless, we had to spend long hours sorting out all our gear and uniforms, setting up our bunks and lockers and generally cleaning up.

A lieutenant involved in the trashing of our barracks took a couple of personal letters I had written to my college sweetheart that had been lying in my locker, stamped and addressed. The next day he called me into his office to tell me that I would have to do extra detail work over the weekend to get them back. I was so upset that I told him to keep them because I knew he could get into trouble for interfering with the U.S. Mail. He forced me to take them back.

One major advantage of AIT over basic training was that after the first few weeks we were regularly eligible for overnight passes. They were called weekend passes, but in reality we were only given Saturday nights off. On Saturday mornings there would be individual inspections of each trainee's living area and gear. If a trainee passed the inspection, he could be rewarded with an overnight pass. With an overnight weekend pass we were on our own; we could even leave the base. All week long I looked forward to the opportunity for a weekend pass.

And now I had friends to socialize with during our time off. One weekend a new friend from my platoon and I got overnight passes. We hitchhiked to Oklahoma City, where we knew no one, simply to escape from the military. We shared a small hotel room, went out to dinner and then went to see the latest popular movie, the science fiction hit *2001: A Space Odyssey*, directed by Stanley Kubrick. The movie was highly entertaining; it introduced the character HAL, the malevolent computer. The movie went on to become a science fiction classic. My friend and I delighted in discussing the movie and space travel in general. An interesting fact is that within a year, while I was fighting in Vietnam, America did in fact land men on the moon. Science fiction became reality.

My friend was a quiet man who experienced the intense fear of the Vietnam War that thousands of young American men were suffering from at that time. He explained to me that he was going to extend his enlistment contract for four more years to guarantee that he would be assigned to an artillery air defense brigade. These were specialized brigades that manned our country's anti-aircraft rocket sites. If he were assigned to an air defense brigade, it meant that he would not have to serve in Vietnam. Since neither the Viet Cong nor the North Vietnamese Army fielded aircraft in the war, our artillery air defense units were not sent to Vietnam. He hoped to be assigned to one of the numerous anti-aircraft rocket sites located around Washington, D.C., where his girlfriend worked.

I challenged his reasoning. Signing up for four more years, a total of six years of active duty Army service, when he had not yet served six months, seemed premature to me, perhaps even foolish. By that time I detested the Army and did not trust the organization. To me, wasting so many years of his young life under authoritarian military rule was a burden beyond endurance. On the other hand, he could not understand how I could take

3. Advanced Individual Training

my chances in Vietnam in order to limit my total time in the military. We lost touch after our training. I do hope that his plan worked and that his dream came true.

Over the course of AIT I developed friendly relationships with a number of other men in my platoon who were from nearby Texas. A few of my new pals were members of a college fraternity that had a chapter house at Oklahoma State University in Stillwater, Oklahoma. One of these fraternity men owned a car that he was allowed to keep on the base. On a number of Saturdays when we were finally awarded our overnight passes, six or seven of us would pile in his car and drive the 130 miles to the university campus. The Oklahoma State University's fraternity brothers were hospitable to all of us Army guys, even though we were not all fraternity members. They allowed us to stay in one of the extra bedrooms in the fraternity house every weekend we appeared. After a few weeks, our stays were so routine that the brothers started calling that extra bedroom the Fort Sill Room.

During our free time on the Oklahoma State University campus we behaved like the college students we all had recently been. We attended football games, went to school dances and drank at fraternity parties. For me, escaping from the burden of the Army's authoritarian rules and regulations was a massive relief. These breaks contributed substantially to my ability to restore and preserve my mental health.

By far my most adventurous weekend escape was the one when a few of my Texas friends and I broke military regulations to attend a college party. The fraternity we regularly stayed at was planning a World War III party, which was basically a military costume party. They invited our AIT group, but on that weekend not all of us received passes. For a reason unknown to me, members of our sister training platoon received overnight passes, but our platoon did not. We were given the same time off, but we were confined to the base. This discrimination was galling because we thought we deserved passes as much as the other trainees, so we huddled together and devised an elaborate plan to remedy the situation.

Our plan was a complicated one. One of my platoon mates was friendly with the trainee who had weekend duty at the brigade's headquarters, where all trainee identification cards were stored. As a favor he agreed to sneak our identification cards out to us on that particular Saturday. These cards were critical because every soldier was required to show his identification card to the gate guards in order to re-enter the base. Also, we predicted that the training sergeants would conduct a bed check in our platoon's barracks that Saturday night. To counter this threat, I convinced an acquaintance in our sister platoon, who had a pass but was not going anywhere, to sleep in my bed on Saturday night. There would be no bed check in his platoon, so he would not be missed. The other men in our fraternity group made similar arrangements with other trainees in our sister platoon.

On that Saturday morning we all passed our inspections and were awarded our specific passes. Later that afternoon we secretly contacted our friend on headquarters duty, who handed our identification cards out to us through a side window of the office. We then piled into our getaway car still wearing our standard olive-green Army fatigues. We escaped Fort Sill without any problem and drove up to Stillwater, arriving in plenty of time to attend the party.

The World War III fraternity party was an elaborate one. The brothers had decorated the walls of the fraternity house's first floor with paintings displaying war themes; for example, one was an atomic bomb mushroom cloud, while others were of tanks and fighter jets. They sandbagged the front door so that everyone had to crawl over that barrier to enter. A few of the fraternity brothers who were in the university's ROTC program had used their connections to borrow real Army jeeps and trucks, which were parked on the front lawn near a civil war cannon that was permanently on display. One fraternity member told me that one brother had planned on borrowing an actual howitzer, so that their fraternity could fire blanks at other fraternity houses, but that plan never materialized.

Of course there was an electric band playing high energy rock and roll music. The party crowd was mainly men and women students from all the university's sororities and fraternities; everyone was required to be dressed in some type of military uniform. Most guests were dancing and everyone consumed their fill of alcoholic beverages. Some of the partiers were young Army officers who taught in the university's ROTC program. We recognized their authentic officer uniforms and avoided them. On the other hand we received multiple compliments for our costumes from students, who did not realize that our "costumes" were so believable because they were in fact authentic Army fatigues. Episodically, sirens screamed out and the band would abruptly stop playing. The sirens announced an air raid, which meant that everyone had to pile into sandbagged bunkers built near the middle of every room. This proved to be great fun as we all tried to jump into the too-small bunkers, piling on top of each other. There were no conflicts because we all were in a fun party mood.

My favorite moment of the evening was exactly at midnight, which denoted the time when we trainees were officially absent without leave

3. Advanced Individual Training

(AWOL). If we had been caught, we would have been arrested and prosecuted. We all had taken a big risk. Our little group gathered on a set of steps a few minutes before 12. When the clock finally struck midnight, we raised our glasses in a toast to our official outlaw status. Then we all enjoyed the rest of the night's party without any incidents or problems. We slept in our Fort Sill Room that night, and the next day we drove back to the real Fort Sill.

We entered the Fort displaying our authentic identification cards. Then we discreetly returned our ID cards to our friend, who was still manning the headquarters building. I looked up my buddy who had slept in my bunk to thank him. He reported that there was a bed check count that Saturday evening, but obviously the DI did not recognize the changes. This positive state of affairs worked out for all of my rebellious outlaw friends. That weekend remains as one of my fondest military memories.

Finally my eight weeks of artillery AIT ended. We did not receive any leave time; instead we were marched to a different section of Fort Sill to begin our new OCS training phase. By that time I was firm in my decision not to pursue the OCS option. My decision was partly due to the fact that OCS meant significantly increased time in my military obligation and partly due to the prospect of six months of extremely stressful training, but mainly my decision was based upon the fact that I would be required to order men into battle in a war that I knew was a huge mistake. It was bad enough that I had to fight in the Vietnam War. I did not need the extra burden of having to order soldiers to do things that could result in their deaths. My subsequent guilt would have been too much to endure. I did not believe in the Vietnam War; my goal was to keep a low profile while I served my time in hell.

Consequently, on the first day of orientation to OCS, I officially made my decision known to the OCS training cadre. They were not pleased, to say the least, but they could not force me to be an officer. Instead they subjected me to a procedure that I suspect is standard for dealing with potential dropouts. Their approach was a deviation of the good-cop/bad-cop police interview. First they put me in a small office alone. Then a soft-spoken friendly sergeant came in to talk gently with me. He attempted to sell me on the advantages of being an officer, for instance the increased pay and status. After I shared with him the reasons for my decision, I detailed my lack of belief in the Vietnam War and how it was not worth the costs and the casualties. He did not have any good counter-arguments other than the need to fight communism, blah, blah, and blah. Our talk only lasted ten minutes before he was stuck.

The nice sergeant left the office and within one minute the bad sergeant entered the room. He was a tall, muscular, angry black man. Even before sitting down he screamed, "You are going to Vietnam, boy!" I answered, "This is 1968. Everyone is going to Vietnam." I knew being commissioned as an artillery officer would not relieve me from a combat tour in the Vietnam War. My answer must have been a good one, because after it this bad sergeant only made half-hearted attempts to frighten me. However, I remained steadfast in my resolve. He left the room after only five minutes.

The cadre finally accepted my decision to drop out of OCS. I was assigned to what was known as a holding company, while my paperwork was processed and new orders cut for me. I knew they would be orders for Vietnam. It took about ten days for the paperwork to be processed. During that time we dropouts—yes there were others like me—were housed in a separate barracks near the OCS barracks. During those days I was ordered to perform various details (military speak for short term jobs) like serving the officer candidates their food. There was no real KP in OCS; the Army hired private contractors to prepare the food, but not to serve it. I found myself serving food to the guys that I had trained with in AIT. They all appeared to be extremely stressed out. One candidate whispered to me that I had made the right decision because they all had been subjected to insanely intense harassment and severe sleep deprivation. After a few days, a few of these same men were standing next to me, serving the OCS candidates their food; they had dropped out too.

Those of us in the holding company were required to attend a morning formation at which sergeants would come to pick men for their various work details. One day I was assigned to guard a laundromat, keeping it clean and ensuring that trainees did not sit on the washing machines. After a couple of days of these types of work assignments, it became obvious to me that they were nothing more than busy work. I decided to avoid them. I teamed up with a sneaky fellow who also wanted to escape these work details. It took applied imagination and plenty of nerve. If I had worried about getting caught, I could not have escaped. In the confusion of all the sergeants picking their men, we would plainly walk out of formation as if we had been picked. Then we would slip away to spend the day at the PX or the library or wherever we could hide, but never in the same place twice.

Our closest call came one day when we hid in an unlocked barracks currently not in use. We were lying down and relaxing in a couple of second-story bunks when we heard the sounds of a small group of officers and sergeants inspecting the building. We could not run without being seen; we

3. Advanced Individual Training

were trapped in the building. Fortunately, we found a storage room where both walls were covered by three-foot-square wooden cubbyholes for storing duffel bags. We climbed up to ones near the ceiling, crawled into the back of two of them and held our breath. We could hear the inspection group walk up to the second floor and move around a bit, but thankfully they never entered the storage room. It was a close call; my heart did not return to its normal rate until I left the building.

Finally my new orders were cut. I was given the standard month's leave before I was ordered to fly to Vietnam. By that time it was the middle of December, and I was able to return to Pittsburgh and spend Christmas with my family. During my leave I shopped with the help of my uncle Joseph Kulakowski and his daughter Mary for an engagement ring. I bought a beautiful three-quarter-carat diamond ring set in white gold. The next day I proposed to my sweetheart and happily she accepted. To celebrate our engagement we flew to North Palm Beach, Florida to stay for two weeks with her older brother and his family. We slept in separate bedrooms but spent the days together enjoying all that Southern Florida has to offer: the warm weather, the beaches, and of course the Atlantic Ocean itself. It was the beginning of our love affair with Florida, which is where we have lived for the last 35 years. We returned to Pittsburgh and far too soon my leave ended.

Sadly, my last day at home arrived. I had packed my duffel bag with my uniforms and olive green dyed underwear and was about to leave when my uncle, Walter Scislowski, knocked on my parents' front door. He was a tough guy who had been a nationally ranked amateur boxer in his youth. He was also an Army World War II combat veteran who had been severely wounded fighting against the Japanese in the Pacific. I gulped when I saw him standing in the doorway. I knew he was showing his affection and concern because he realized how horrible it was to fight in a war. I greatly appreciated his taking the time to see me off. My fear of the Vietnam War expanded the moment I laid eyes on him that morning.

4

Arriving in Country

My Army orders read that I was to fly to Oakland, California, on January 22, 1969, to await air-transportation to the Republic of South Vietnam. I flew from Pittsburgh on a commercial jet liner at the government's expense, arriving in the early evening. At the Oakland military terminal there were no sleeping facilities; consequently enlisted men were harassed with busy work all night. I was ordered to count boots in a gigantic warehouse half filled by a mountain of used Army boots.

Because the terminal had little in the way of perimeter security, more than a few men chose to go AWOL to travel to the city of Oakland or onward to San Francisco to party during their last night on United States soil. After some consideration of this option, I rejected the escape because I recognized that after a night of partying, the temptation never to return would be intense. Of course, I had no real desire to fight in Vietnam, but to go permanently AWOL at that point in my enlistment would be viewed as cowardly even by my own family. My bottom line was that I was not going to be a cowardly embarrassment to myself or to my family.

Even though we were far from a military family, when America called them to arms, the men of my family had responded courageously. During the last year of World War II my father, Walter Francis Senior, had hunted for German submarines while stationed on a destroyer escorting supply convoys to England on the stormy North Atlantic Ocean. My Uncle Frank Kulakowski had fought the German Nazis for years as an Army infantryman in Europe. My Uncle Walter Scislowski, who had boxed in the Golden Glove national tournament, fought as an Army combat engineer against the Japanese in the Pacific. He captured the rifle of a Japanese soldier that he had killed. This war trophy rifle hung on one basement wall of his home, where as a boy I admired it with wonder. Two of my cousins had fought

4. Arriving in Country

bravely in the Korean War as Marine riflemen. My attitude was that now it was my turn to fight for my country.

However, a critical question for any war is this one: What is the soldier's perception of the legitimacy of his war? I know that some Israeli soldiers during the 1973 Yom Kippur War could look back over their shoulders in the Golan Heights and actually see their homes. Those Israeli soldiers knew what they were fighting for. Soon after the Pearl Harbor attack, American soldiers were constructing machine guns nests on California's beaches in anticipation of a feared Japanese invasion. In stark contrast, while fighting in South Vietnam, not once did I fear that the North Vietnamese or the Viet Cong were going to jump in their sampans and row across the Pacific Ocean to attack California. I knew we were fighting Communism, but to me the Republic of South Vietnam, a corrupt aristocracy governed by the rich few, was not the best place to draw the line. Was the South Vietnamese Republic really worth the high price we were paying in thousands of American deaths and billions of dollars spent? My opinion was and still is: No. Yet a soldier is not given a choice of wars. I was being sent to fight in the Vietnam War no matter how much I questioned its legitimacy. Plus, there was no real future in taking the AWOL escape route. I had plans for my future.

So here I was counting used Army boots under the watchful eye of a duty sergeant. Thankfully, the task was so boring that he eventually drifted out of the warehouse. I took advantage of the opportunity to bury myself within the pile of boots. I tried to sleep, but I achieved only limited rest because running through my mind throughout the night was my new reality: "I am on my way to WAR."

The next morning I boarded a commercial American jumbo jet liner. It was a surreal experience, flying off to war on a luxurious jet liner crewed by young attractive stewardesses, who pampered and fed us like we were all first class passengers. They were cheerful and attentive, like it was perfectly normal to serve a plane full of young soldiers dressed alike in their khaki uniforms—except that no one was flirting with them. All the men's faces were tense with apprehension, if not pure dread.

First we flew to Alaska, landing there, I presume, for refueling. Then we flew directly to South Vietnam. The tension among us grew over the hours of the flight, especially as we neared our destination. Finally, on January 23, 1969, I landed at Tan Son Nhut airport located on a large Air Force base in the northwestern outskirts of Saigon. Another surreal moment was when the pilot spoke over the intercom as we deplaned: "Welcome to Saigon, local time 2 p.m., temperature 98 degrees, 100 percent humidity, light ground fire. Thank you for flying American." Two stewardesses stood warily on the inside of the exit doors, repeating over and over their farewell of, "See you next year," reflecting an optimism that each one of us would be returning after the end of our one-year combat tours. I hoped they were right.

The pilot had been accurate. As I walked down the steps that had been rolled up to the plane's door, I had my first taste of South Vietnam's intense heat and humidity. It was the worst weather I had ever experienced; it felt like I was suspended over a large pot of boiling water. It was a moist heat that clung to your skin. At the bottom of the stairs I heard warning sirens start to blare and the thump of distant explosions. Welcome to Vietnam! The air base was undergoing an enemy mortar attack.

There was no one on the tarmac to greet us or tell us what to do. We were on our own. Exploding mortars continued to hit the air base, but I estimated by their sound that the mortar shells were not falling nearby. Tan Son Nhut was the largest air base in the country and this appeared to be a limited attack. If the Viet Cong were trying to destroy the incoming commercial jets, they were failing. If their goal was to frighten the newly arrived American soldiers, they were a huge success. I ran to a nearby large truck and hid behind the front fender. It was a good hide from which I could evaluate the situation.

I kept my head up and observed that the mortar shells were not striking any closer; however, to my exasperation, I saw that many of my fellow new arrivals appeared to be in full-blown panic. I saw soldiers literally running around without plan or reason. They would run to one spot, hide for a second or two and then run to another hiding spot. Then after a few more seconds they would run to another hide in a desperate search for absolute safety. Real safety in the form of reinforced sandbagged bunkers was not available.

My assessment was that hunkering down in one reasonably protective spot was the safest strategy. Later in my combat tour I learned that my assessment was the correct one. The worst thing for a soldier to do when under a mortar attack that causes hundreds of pieces of razor-sharp shrapnel to whizz through the air is to be upright and above ground. As the distant mortar fire continued with its accompanying explosions, I thought to myself, "There is a lesson to be learned here. When under fire, Do Not Panic! Use your head and try to figure out what exactly is going on."

Eventually the sirens and the mortar attack ceased. Sergeants finally appeared and gathered us up to begin the lengthy administrative processing into the Vietnam War zone.

5

Tan An

Tan An, a small town 15 miles southeast of Saigon in Vietnam's Mekong Delta region, was the site of my first official duty station in the war. It was the headquarters of the 2nd of the 4th Artillery Brigade that had been attached to the 9th Infantry Division. Of course, the Army did not assign me to do the military occupation I had been trained for, fire direction and control. Instead I was ordered to monitor the military radio traffic of infantry companies embroiled in battles with the Viet Cong or the NVA and to communicate with the infantry units as needed. My duty shifts were 12 hours long.

It was not demanding work, but it was a stressful place to be stationed. Besides trying to eat and sleep during the 12 off-duty hours, my extra time was filled with work details, many of which seemed to be busywork. For example, a group of us enlisted men were ordered to sweep unused buildings or change the oil in jeeps more frequently than the factory-recommended intervals. The brigade's second in command—known in military speak as the executive officer—was a major who developed a negative opinion of me, most likely because I had difficulties hiding my disgust with him, his busywork details, and the entire Vietnam War. Even today my wife informs me that I cannot hide my feelings. My emotions are always on display on my face and in my body language. Consequently, I was always assigned to these after-duty-hours work crews. The major appeared to relish the hours that he spent personally supervising our vital contributions to the war effort, like picking up trash. Apparently he had nothing more important to do. I worried about losing control and punching the major.

Like all low ranking enlisted men on the base, I was assigned living quarters in an eight-man tent. Sandbags were stacked up four feet high along the outside walls, supposedly to protect the tent. In reality these stacks of sandbags provided little real protection from enemy attacks. I was

Me smiling during the first month of my combat tour at our brigade's headquarters in Tan An.

issued a wooden cot and a metal locker. There were no dividers between bunks, and therefore no privacy. We all slept in a horseshoe pattern of cots along the tent's walls.

The Tan An headquarters base was primarily devoted to command and administration of the brigade's various outlying artillery batteries located at firebases scattered across Vietnam's Mekong Delta. The base did not have its own howitzers; instead it had only four 80-millimeter mortars for defense. Assigned to the base was a small airport for helicopters and light observation airplanes. It was located about three miles away from the base at the edge of the town. The plan for the base's defense relied mainly on the brigade's artillery located at other firebases to fire on enemy attackers outside our perimeter, and also to utilize the helicopter gunships stationed at our small airport to beat back enemy attacks.

Whenever any of the 9th Division's infantry companies was caught in a firefight with the communist enemies, I was ordered to keep a running tally of American casualties, both wounded and killed—the acronyms were WIA, wounded in action, and KIA, killed in action—as well as enemy KIAs and any captured enemy weapons. It may sound like an easy task, but it was actually a difficult one because the soldiers I talked with on the radios were always so stressed during the actual fighting that their reports were confused, inaccurate or delayed. What made my job especially difficult was the officers standing behind me yelling continually, "What is the body count? How many enemy bodies do they have?" In their frenzy they could not wait until the firefight was over. They expected the combatants to count and fight. I resisted their commands as best I could.

Enemy body counts had become entrenched as the ultimate criterion for combat success early in the Vietnam War. It was a criterion that the commanders could quantify. This is how the American military establishment evaluated its military success against the Vietnamese communists, even though their leader, Ho Chi Minh, had been quoted as stating, "You can kill ten of our men for every one we kill of yours. But even at those odds, you will lose and we will win." The Vietnamese communists repeatedly demonstrated that they were willing to suffer massive losses to win the war. Plus, they could send North Vietnamese reinforcements down the Ho Chi Minh Trail and forcibly conscript South Vietnamese military-aged men faster than we could kill them.

But these realities did not seem to matter. The officers were all crazed over the concept, always yelling at me for the body count, even if the battle had just begun. One time while I was monitoring a firefight, the S2 major actually tore my radio's microphone out of my hands to demand body counts from the soldier I had been communicating with. That soldier angrily responded that he was too busy fighting for his life to be counting enemy bodies. His justifiable plea did not faze the major; he continued to shout into the microphone, "I want my body count, I want my body count." I dared not say a thing during this pathological encounter, but I think my revulsion showed.

All my negative emotions were on display. These were the first weeks of my direct involvement in the real war. To have to tally the numbers of Americans killed or wounded fighting the communists upset me. These infantry firefights were ugly, frightening and depressing. I was counting American soldiers brutally killed or horribly wounded; traumatic amputations of our soldiers' arms and legs were far too common. Also, many of these battles ended with multiple dead Americans but few, if any, communists' bodies. Usually all that was left of the enemy were blood trails as the communists pulled out their dead and wounded. The crazed officers demanding body counts added to my horror. The entire exercise disgusted me. I became seriously depressed and tried to exist in a state of semi-numbness. I was afraid that I would drown in all the ugliness. In my February 1 letter to my sweetheart, I wrote, "It's hell being a young man in this day and age." It was a hateful time.

After about four weeks at my new duties, I awoke one night in my tent screaming out in fear. My nightmare was that of a vicious yellow face growing gigantic as it charged at me. When I fully awoke, I was sitting up in my bunk gasping. I then became embarrassed by my outburst of shrieking fear, but as I scanned the dark tent, I was relieved to see that I had not awakened any of the other seven men. My actual scream was apparently not as loud as it sounded inside my head. Thankful that I would not be branded as a frightened newbie, I exhaled, yet my heart was still racing. It took a couple of minutes before I was able to fully calm down. Totally ashamed, I vowed to myself never to be so terror stricken again. It worked; I never again was afflicted by a nightmare during my entire yearlong combat tour.

My most hazardous duty at Tan An was acting as the shotgun guard on a jeep. For various reasons small aircraft would land at the Tan An airport late into the night. A driver would then be ordered to drive out to the airport to pick up the pilot. A number of times I was ordered to be their armed guard. At that time I had not been issued a flak jacket or a helmet, so there I was in the middle of the night sitting up on the back of a jeep in my jungle fatigues and a bush hat, armed only with my M-16 automatic rifle. Warily I scanned the eerie darkness as we drove through the unlit town and down the lonely road to the airport. Usually there was no other traffic that late, which added to the menace. We were the sitting ducks at a shooting gallery, vulnerable to any rifle fire, especially from a trained sniper. The driver and I never talked; we were both too anxiously vigilant. I was always pondering over how best to react if we were fired upon. Fortunately we were never attacked on my guard duty trips, but these dangerous runs significantly increased my generalized anxiety.

5. Tan An

It was at Tan An that I suffered through my first enemy ground attack. The VC had totally surprised the American military when they launched their 1968 Tet Offensive. So in early 1969 there was much apprehension and corresponding defensive preparations as Tet, the Vietnamese New Year celebration, approached. The communists did launch an offensive, but it was scaled down significantly from the previous year's attacks. One night we had just turned out the lights and were preparing to sleep when the alarm siren wailed. A staff sergeant ran into our tent to announce that the VC were attacking our perimeter. He ordered us all to dress, grab our rifles and ammunition, and head out to defend our berm. As we eight were running out of our tent, the staff sergeant, for whatever reason, grabbed me and ordered me to guard the entrance to the brigade's command bunker, known as the Tactical Operations Center (TOC). He told me that one of my duties was to ensure that all of the officers entering the TOC unloaded and cleared their weapons. Our commander did not want any officer in the TOC shot by an accidental weapons discharge.

I dutifully ran to the TOC and set myself up as the entrance guard. Soon a number of high-ranking officers ran up to the entrance. As ordered, I told each one to take a moment to clear and unload his weapon. Most were armed only with the standard Army-issued sidearm of the time, the 1911 .45-caliber semi-automatic pistol. They all complied without a complaint, acknowledging the wisdom behind the order.

After the officers entered the TOC bunker, I was left alone and frightened. Being caught in my first real battle was a nerve-wracking experience. I could hear the gunfire at the berm and see the enemy's green tracer rounds flashing over the base. From one of the officers I heard that we had already lost nine soldiers, killed by the enemy's opening salvos. We were not as prepared for the battle as we should have been. Our commanders had not cut back the thick jungle foliage that had grown up to the edge of the perimeter. This enabled the VC to climb high enough into nearby trees that they were able to shoot down directly into the sleeping quarters tents. The walls of these tents, like ours, were not sandbagged high enough. The VC's AK-47 bullets had easily pierced one tent's canvas walls, killing nine of the sleeping soldiers inside.

The battle raged on as I stood my lonely guard duty. I reminded myself of the importance of not panicking in battle. I forced my mind to formulate my own personal battle plan. If the VC attack broke through, from what direction would they most likely charge at me? Where would be the safest position to fire on them? I imagined the enemy attacking me and me shooting back at them. Thinking up my battle plan strengthened my resolve.

This mental preparation calmed me down a little. I felt more prepared, even if I were wrong about the direction of the attack. I had my M-16 set for fully automatic fire; I had a round in the chamber; I was determined to shoot any enemy I saw. I was ready to fight; thankfully I did not have to. I never saw the enemy.

At first, only our base mortars were firing at targets just outside our berm. Soon heavy artillery fire from supporting firebases also started hitting near our berm. After a few minutes the heavy weapons stopped as the Cobra gunships arrived on the scene. The Bell Company's Huey Cobra gunship was the first specifically designed fighter helicopter. Its defining features were its speed from the turbocharged engine and its width. The Cobra's fuselage was only three feet wide, just wide enough for the pilot to fit in with his gunner/copilot sitting directly behind him. Its narrow profile made the Cobra difficult to hit as it flew attack runs directly at the enemy. Cobras had two 7.62-mm multi-barreled miniguns mounted into the nose and a rocket pod under each of the stubby wings. Each rocket pod held 19 70-mm rockets.

Paired with each Cobra, in what were known as "hunter-killer teams," was an unarmed light observation helicopter (LOH) nicknamed the "Loach." In our battle the Loaches were crewed by a pilot and a copilot manning a powerful searchlight. They recognized that the VC fire was coming from the tops of nearby trees. I saw one Loach searching tree to tree with its spotlight, looking for the VC snipers. When an enemy was spotted, the Loach would hover near the tree and shine its light directly on the enemy. Then its Cobra partner would make a high-speed gun run directly at the VC sniper tied to the tree. Dropping in altitude while blasting away, the Cobra would swoop past the constantly hovering Loach by just a few feet. It was quite the courageous tactic. Sometimes the Loach was so close to the trees that it seemed to me that an enemy fighter could jump into the light helicopter and kill the pilot with only a knife. Thankfully, all the enemy soldiers spotted by the Loaches were blasted away into oblivion.

After about an hour the battle wound down, the all clear signal was sounded, and we were allowed to return to our tents. My tent mates were still hyped up from the firefight. They bragged how they had defended a portion of our base's defensive perimeter from a direct VC ground assault. Their rifle and machine gun fire had stopped the attack. One man had actually stood on the top of the mound and shot downward with his M-60 machine gun at the VC attackers. He was still excited about the fight, as he had a right to be. I have to admit that I was a little jealous of these

5. Tan An

Side view of Cobra helicopter gunship showing rocket pods on stubby wings as well as rapid fire mini-gun under its nose.

men who actually saw the attacking enemy and had a chance to shoot them. I wanted to fight back too.

These young artillerymen, who had valiantly defended our berm, were later awarded the Bronze Star Medal for their valor. I was awarded the less prestigious Army Commendation Medal, yet I am still proud of that award. I had been involved in my first direct ground attack and had done my duty. I was proud of myself for not panicking. Most importantly, we had killed VC, beaten back the attack, and suffered no more American deaths after their initial onslaught. It smelled like victory.

At the end March brigade headquarters learned that replacements were needed for two jobs at the brigade's frontline firebase near the Tan Tru village. The major, who disliked me, considered these openings as the perfect chance to get rid of me. One of the replacement positions was for a radio/telephone operator (RTO) for a lieutenant forward observer (FO) assigned to one of the firebase's infantry companies. The RTO would accompany the FO on all of the infantry's combat patrols. It was almost as if the RTO were an infantryman, a dangerous job indeed. The major wanted to assign me to this RTO position as a form of punishment.

Fortunately, an Irish-American first sergeant took pity on me because

I too was of Irish descent and not that terrible of a soldier. He saved me by convincing the S-2 major that I would be a better fit in the other replacement position, that of a liaison specialist, a job that involved coordinating artillery fire for the base's infantry battalion. The soldier who held the job before had been a RTO and had been brought in from the field after seven months of front line combat duty. The problem was that the liaison job proved to be too complicated for the man. He never fully mastered all of his duties, causing officers to worry that he was going to make a mistake that would kill some of our soldiers. Consequently they removed him from the position and were requesting a bright artilleryman who could come in and quickly master the job. I was that soldier.

After I packed up my rifle, uniforms and personal items, I was immediately driven by jeep the seven miles northeast to the artillery firebase built next to the village of Tan Tru. My anxiety grew on that drive because I knew that I was on my way to the Vietnam War's version of the front lines, an infantry battalion's base camp combined with two artillery batteries, known in the war as a "firebase." Here was a base and a job where I would be more directly involved in combat operations than I had been at Tan An. I was not exactly sure what I would be getting into.

The buck sergeant who led the artillery liaison crew turned out to be a nice guy. He welcomed me warmly to the firebase and he helped me set up my bunk, which was in a large above-ground bunker. It housed only the battalion's headquarters enlisted staff and snipers. The bunker's outside walls were completely lined with the heavy wooden crates that artillery rounds were transported in. After the rounds were delivered, the crates were filled with dirt so that they could perform the same protective function that sand bags did. The buck sergeant guided me through the many introductions to the staff officers and then taught me the specifics of the job. At first he actually demonstrated what needed to be done, and then he supervised my beginning attempts to do the job on my own. As a motivated student, I quickly acquired the necessary knowledge and skills because I did not want to end up being an FO's combat RTO. This pleased the friendly buck sergeant. He was relieved to have someone in the critical position who could completely shoulder the job's many serious responsibilities.

Even though I was now in a more direct combat-oriented job, I came to like the position of liaison specialist. The job's major requirement was to double-check artillery targets to prevent friendly fire: that is, to prevent our artillery from killing our own soldiers or any of our allies or Vietnamese civilians. I was and still am proud of the fact that my primary military function in the Vietnam War was to save lives rather than to take them. An added bonus was that there were three enlisted men who shared the position, so our shifts were only eight hours long. Since I was the new guy, I was assigned the night shift, from midnight to eight in the morning.

6

Liaison Specialist and a Failure

As previously mentioned, my military occupation specialty training was in artillery fire direction and control (FDC), which is aiming artillery howitzers, but again I was not assigned that job at the Tan Tru firebase. My new official Army job title was Liaison Specialist, Fourth Class. Administratively, I was classified as an artilleryman, but I worked and lived with the infantry. Our liaison team consisted of an artillery captain and three enlisted men; our job was to coordinate artillery fire for the infantry battalion in which we were embedded.

I was now stationed at a firebase near the Long An province village of Tan Tru, located about 15 miles southwest of Saigon in South Vietnam's Mekong Delta.

Our firebase consisted of one infantry battalion, the 2nd Battalion, 60th Regiment of the 9th Infantry Division. It consisted of four infantry line companies of approximately 100 men each, a reconnaissance platoon of 20 men, and support units like supply and the motor pool. Our firebase also included two artillery batteries. One battery was comprised of six 105-millimeter howitzers; the other was comprised of six 155-millimeter self-propelled howitzers. Both batteries were part of the 2nd Battalion, 4th Artillery of the 2nd Field Force.

The 105-millimeter howitzer is the most common sized artillery piece in the United States Army. It fires a 32-pound shell that can reach a maximum range of seven miles. The shell's warhead explodes on impact. The howitzer is a more versatile weapon than a cannon because its barrel can be elevated to fire at a high trajectory, giving the howitzer the ability to strike targets located behind large obstacles.

Our firebase's six 155-millimeter howitzers were self-propelled tracked

6. Liaison Specialist and a Failure

An aerial view of the front entrance to our Tan Tru firebase.

vehicles that looked like large tanks; however, their turrets could not turn like those of tanks. The 155-millimeter howitzer is truly a devastating weapon. It shoots a 95-pound shell with an exploding warhead that can reach a maximum range of nine miles. When a 155-millimeter steel-cased warhead crashes into a target, it explodes, producing a killing zone 110 yards in diameter. Image that lethal zone, larger than a football field, where every unprotected living thing would be exposed to hundreds of red-hot, razor-sharp steel fragments whizzing through the air at extremely high velocities. Any one of those steel fragments could kill a human. Even if a soldier were heavily protected, if he were near, the blast's shock wave concussion alone would kill him.

We also had under our control our infantry battalion's mortar platoon, which consisted of both 81-millimeter mortars and the more powerful 4.2-inch mortars. The 81-millimeter mortar fires a 15-pound explosive projectile out to its maximum range of 3,200 yards. The 4.2-inch (four-deuce) mortar fires an 11-pound explosive projectile out to its maximum range of 4,400 yards. At our firebase the mortars were mainly utilized as short-range defensive weapons.

My primary mission as an artillery liaison specialist was to draw an accurate map of friendly military units in our battalion's area of operation (AO) and to ensure that our artillery did not fire on these friendly forces

Typical 105-mm howitzer gun pit with crew at our Tan Tru firebase.

or civilian population centers. In other words, my job was to prevent what is termed "friendly fire." In our AO we had the American Army, an elite South Korean Army Division, and the Army of the Republic of Vietnam battalions all patrolling in a region with a dense civilian population. There were also American Navy forces (known as the Brown Water Navy) patrolling over the Mekong Delta's numerous inland waters.

When one of our artillery FOs attached to an infantry company wanted to fire on a specific target, he would radio the target's map coordinates into the FDC bunker located near the howitzers. FDC would then begin computing the firing solutions and relay the target's coordinates to me via radio. I would plot the new target's coordinates on the large up-to-date AO map pinned to one wall of our TOC's radio room. If the target were not near any of our maneuvering friendly forces or near a civilian village or town, I would then provide FDC with my initials, which were then termed the safety clearance initials. Even to this day I am proud to report that I never made an error, that Tan Tru's 2nd/4th artillery batteries never caused injury or death to one American soldier or any of our Allies' soldiers during my combat tour in South Vietnam.

Besides on-call target safety clearances for troops in the field, potential artillery targets could also be pre-cleared. Every day at least one of our

6. Liaison Specialist and a Failure   47

battalion's four infantry companies would be out in the field on patrol, searching for the Viet Cong or the North Vietnamese. The exact location for each patrol was determined by our battalion's commanding officer, a lieutenant colonel. During what were called "jitterbug" operations, the colonel would select, based upon various forms of intelligence information, dozens of potential targets for small unit airmobile reconnaissance. Each night I would receive the grid coordinates for up to 50 insertion targets for the next day's jitterbug. Then my job was to plot these targets on my up-to-date large-scale AO map. If any of the targets were within 1,000 yards of an occupied village, I had to contact the village chief, who was the only person empowered to grant us permission to fire at the nearby target. The rule was that each potential target include a 1,000-yard buffer zone around it. Even if only a small corner of a village was within the edge of that buffer zone, I was required to seek the chief's approval.

I was also required to contact any other military units who were scheduled to operate in or near our AO that day. We would exchange our respective jitterbug target coordinates and then plot them to ensure that there would be no contact or friendly fire between our patrols. Only those targets that were outside all buffer zones would be approved and given my safety clearance initials. My counterparts stationed at other firebases in our region would do the same for our battalion's set of potential targets.

The next morning the colonel would prioritize a smaller set of targets out of the entire list of cleared potential targets based upon new intelligence or his intuition. This new set of targets would be the actual targets the infantry would search that day. Hueys would airlift an infantry company to the first target area. The infantry would search for signs of the enemy while the colonel and his staff officers would direct the search as they flew overhead in the Command and Control (C&C) helicopter.

As my on-the-job experience grew, I began to question, "Was this our best military master plan?" Our intelligence information on the enemy's location was rarely accurate, so target selection depended mainly on the colonel's intuition. His choices were usually wrong. Most of the time the infantrymen did not find the enemy. Usually, after a reasonable amount of time, the colonel would decide to end the search in order to try the next target on his prioritized list. He would order in the lift helicopters and the infantry company would be airlifted to another target area. In fact, one day our infantry battalion hit 16 different targets, swept each and came up empty each time.

The airmobile operations became a matter of chance, just simply probing for a reaction, like someone stabbing at possible hives trying to excite the bees. Besides, one enemy sniper shot could provoke the colonel to call in more troops for reinforcements. Thus one sniper would tie down entire companies of American soldiers. And if there were an American casualty, the infantry companies could be immobile at times for hours as the medical evacuation helicopters were called in and the wounded soldier was airlifted to the closest military hospital. It became apparent to me that jitterbugging was not an effective guerrilla war strategy.

Huey resupplying our infantry in the field.

I proudly mentioned that I never lost a soldier to our own howitzer fire. That is a true statement; unfortunately, however, there was one tragic incident in which Vietnamese civilians were killed because of an artillery officer.

One night our inexperienced battalion FO was flying with the colonel in the C&C helicopter over one of our infantry companies, which was engaged in a firefight with the VC. I remember him because he was the only one of my four different captain bosses who had been commissioned as an officer after graduating from artillery officer candidate's school. All the other captains I worked with were West Point graduates. He called in six targets for our 105-millimeter howitzers to fire on. As I heard the targets over the radio, I automatically plotted them on my AO map. Five of his targets were near each other in the general area of the firefight, but the last one was an outlier, a good distance away from the other targets. What probably happened in the confusion and stress of battle was that he transposed two numbers of that outlier target's grid coordinates. For example, instead of calling in his intended target at grid coordinates 685429, he radioed in orders for grid coordinates 658429. It was obvious to me that the one target was probably a mistake.

The artillery RTO on the ground with the infantry company and the battery's FDC specialist also spotted the discrepancy. All three of us communicated our questions and concerns to the artillery FO captain in the C&C helicopter. He never really considered our concerns or double-checked his map; instead, he pulled rank. Immediately he radioed back:

6. Liaison Specialist and a Failure

"I am up here, you are down there.
I am an officer, you are enlisted men.
Follow my commands. Fire on those targets!"

After such a direct order from an officer, there was little we could do but obey. I double checked the targets, and by sheer chance the outlier target happened to fall in the buffer zone of a cleared target. Remember, I routinely cleared dozens of potential targets before any of our jitterbug operations. That day one of those targets was close enough to a Vietnamese village that its 1,000-yard buffer included a corner of the village. Possibly the village chief had not studied each individual target closely. Anyway, he had given me his safety clearance initials, which in effect permitted us to fire on that nearby target. As protocol dictated, I logged that target's coordinates and the chief's clearance initials in my daily ledger of cleared targets. Next I gave our howitzers the clearance initials for all six targets.

Soon afterwards our 105-millimeter howitzers fired on all six targets, including the questionable one. One of those lethal howitzer rounds crashed into the corner of that Vietnamese village, killing 14 civilians and four water buffalo. Instantly someone somewhere ordered a cease-fire.

A few weeks later I was unexpectedly called out of my bunk at ten in the evening to report in full uniform to the TOC. Upon arriving, I learned that there was an ongoing military justice investigation into the tragedy. One of the investigating officers, a major whom I had never seen before, ordered me to stand at attention before him while he questioned me about the details of the incident. He ordered me to report step-by-step the process involved in clearing that target, which I did, including my communication of concern to the FO about the outlier target. The inspector's manner and style of questioning was so accusatory that I stopped at one point and spoke up. I asked if I, the safety clearance specialist, was being accused of misconduct. The major backed off a little and reported that the FO captain, who had ordered the strike, had admitted his mistake, accepting full responsibility for the incident.

My defense was solidified when, after a brief search, I was able to produce in our Safety Clearance Ledger the coordinates of the target under investigation. This official ledger record was dated and the line after that target included the village chief's clearance initials. On my large AO map I was able to demonstrate to the major that a corner of the village did indeed fall into the 1,000-yard buffer zone of the previously cleared target.

The inspecting major ordered me to submit a written statement about the incident and then dismissed me without any discussion of the reality that the tragedy could have been prevented if the FO captain had simply considered my concerns and double-checked his maps. Instead, this FO captain chose to pull rank. When I wrote up my statement later, I included the fact of the captain's arrogance and submitted it. The inspecting major did not accept my statement; instead, he intimidated me into changing it. Knowing full well how he could ruin my life, I finally agreed to have the major write up what he wanted and I would sign it. The final document did not blame me, but it did portray the captain in a better light. Two weeks later I found out that the arrogant artillery officer's punishment was nothing more than an administrative one, in effect a slap on the wrist. I also learned that the U.S. Military Command in Vietnam paid $40 to the Vietnamese families for each villager that had been killed in the artillery strike. The rice farmers were paid $70 for each of the water buffalo killed.

I was shocked by the news. I simply could not decipher what kind of upside down, never-never-land this was here in Vietnam, where water buffalo were considered more valuable than human beings. To me this stupid tragedy revealed the inherent weakness in the rigid hierarchical system of authority that was our American Vietnam Era Military. Bottom-up observations or information from enlisted men was discounted or, as in this case, completely ignored, under the false assumption that officers could learn nothing from the lower ranks. Fourteen civilians died here needlessly because of the Army's rigid authoritarian system.

7

Air Power Support

One day during the 1969 dry season I was ordered into the field to act as the battalion FO's radio telephone operator (RTO) during one of our battalion's doughnut operations. As usual, I was airlifted out to the battle site, which was in abandoned rice paddies located in a remote northwest section of our battalion's AO. On this insertion a half-dozen headquarters company support soldiers accompanied me. Usually these enlisted men worked as jeep drivers or clerks. Like myself they were not involved in the day-in and day-out routine combat operations, but were called upon to add their rifles to these types of seal and pile on operations. We landed uneventfully near the colonel and his small group of support officers, who had already set up a command center on the edge of a rice paddy dike.

What was unusual about this particular operation was that our battalion was depending upon airpower for its heavy firepower support. We were relying upon F-4 Phantom II fighter/bombers flying out of the Tan Son Nhut airbase west of Saigon. My standard procedure for air support was that, instead of direct communication with the pilots of the attacking jets, I was to communicate with an intermediary who was an air-observer pilot flying a Piper Cub–type single seat, propeller-powered airplane. I communicated our battalion's firepower needs to this air-observer officer, who then made the command decisions about which planes and which types of ordnance would be dropped on our enemy targets. The weapons that he chose from included an array of high-explosive bombs, rockets, and the gruesome napalm canisters.

The air observer's Piper Cub plane was vulnerable to enemy ground fire because it flew low and slow over enemy positions. Early in the war the VC and the NVA would direct heavy anti-aircraft fire on these Piper Cub observation airplanes. Sadly, a number were shot down and their pilots killed. Yet even though these small engine planes carried no weapons or armor, our enemies eventually learned how lethal they could become. If the enemy's initial anti-aircraft burst of fire did not bring down the Piper Cub, the communists were in serious jeopardy. Shooting at a Piper Cub revealed their anti-aircraft weapons' position to the Cub's air-observer pilot. He would then know exactly where to direct his F-4 fighter/bombers' attacks to destroy the enemy's anti-aircraft guns. Destruction of anti-aircraft weapon sites became so frequent that communist commanders stopped firing on these tempting observer airplanes.

In this particular combat operation, we were hunkered down behind a rice paddy mound when the air observer spotted a large VC patrol moving up on our position for a direct attack. He warned us of their approach and then ordered a sortie of two F-4s to drop napalm on them, because the enemy was too close to our position to use high-explosive bombs. He ordered the fast movers, a nickname for our jet fighters, to make their bomb runs parallel to our defense lines to prevent an accidental drop on us. Napalm bombs are large metal canisters tapered at both ends. A particularly vicious weapon, the canisters contain a highly flammable, almost explosive, formula of gasoline mixed with a thickening agent that turns the gasoline into a thick jell. This deadly jell would stick to and incinerate anything or anyone it touched.

The attacking F-4s screamed low across our horizon, dropping the silver napalm canisters only about 150 yards in front of our paddy dike perimeter. The canisters tumbled down, hit the ground and exploded into huge fireballs over 50 yards long and 30 yards high. The fireballs were truly impressive, blistering the jungle and anything in it. I was astounded by the size, intensity and horror of the inferno. There was no cheering or celebration among us soldiers huddling behind the dike wall. We were all, I think, secretly thankful that we were not the victims of this horrific weapon. I heard eerie sounds coming from within the fireball; they may have been the screams of the dying or echoes of their screams or the rush of air being sucked into the flames. Not only could napalm burn a soldier to death, but it could also asphyxiate soldiers by sucking the oxygen out of the air near the fireball. I stared in amazement as the inferno cremated the jungle; we received no more enemy rifle fire from that area.

Yet the fight was not over. The communists had maneuvered a 51-caliber heavy machine gun to a position where they could fire the weapon at the F-4 fighter/bombers when they were most vulnerable. Our fighter jets were vulnerable at the end of their bomb runs, when they lost speed as they throttled up and slowly gained altitude.

The enemy's heavy machine gun opened fire. Listening on my FM radio I heard the air-observation pilot scream out in anger, "I'll teach those bastards to fire on my airplanes!"

He must have called for every available fighter jet from the Tan Son Nhut airbase. Soon sortie after sortie of two F-4 fighter/bombers arrived at our location to bomb the communist soldiers in front of our lines. After a time the sorties started stacking up.

Looking skyward I could see at least eight pairs of F-4s circling overhead at various altitudes, stacked like pancakes, waiting their turn to attack the Viet Cong. The air above looked like Chicago's O'Hare airport during the busy Christmas season, when jet airliners would stack up and be forced to circle the airport waiting for an open runway. Each fighter jet made two attack runs; the first was either dropping high explosive bombs or napalm canisters and the second was a gun run with the jet's rapid-fire cannons. The fighter/bomber attacks went on and on for at least two hours. All Viet Cong incoming fire ceased. Some of our officers even dared to stand up.

Our battle was close enough to Tan Son Nhut airbase that the fighter jets could expend their ordnance, fly back to the airbase, reload and refuel to return for another attack. I learned from another air observer that fighter pilots liked to burn off excess fuel before returning to base because it was safer to land with a nearly empty fuel tank. Also, if they returned late enough they would not have to rearm and fly again that day. Apparently some of the jet pilots were becoming fatigued. Late in the afternoon I heard one pilot calling out a warning on my FM radio: "Keep your head down." Soon afterwards his F-4 Phantom jet came screaming down from 5,000 feet to about 30 feet above our position within seconds.

It was a thoroughly unnerving experience. A 30,000-ton, two-engine F-4 fighter, that can easily fly over 1,000 miles an hour, looks like an enormous bomb when it is falling out of the sky aimed directly at you. None of us were sure that the jet would be able to pull up and out in time. All those standing officers dove to the ground. As the Phantom screamed over us, it looked like the jet almost clipped the tops of our radio antennas. Then the jet's eardrum-bursting sonic boom pounded our ears. I imagined those pilots—the F-4 is a two-seater aircraft—laughing at us frightened land dwellers, while our officers lying on the ground cussed the pilots loudly and repeatedly.

8

Dinner with a Wealthy Vietnamese Farmer

Our infantry battalion's AO was Long An province, located roughly 20 miles southwest of Saigon. It was a region densely populated by small villages of peasants, who farmed the province's fertile rice paddies.

There were also a few compounds of wealthy farmers, who owned many acres of rice paddy fields. During the dry season of 1969, new intelligence informed our colonel of recent Viet Cong activity in one of these areas of large rice farms.

In response, our commanding colonel ordered two of his infantry companies out on search and destroy missions within this area. He also ordered headquarters company personnel, soldiers who usually provided support by working in such jobs as drivers or clerks, to the field to help reinforce the two companies' search and destroy patrols. As a member of the infantry headquarters company, I soon found myself on foot, patrolling over endless rice paddy dikes in the sweltering Mekong Delta heat and humidity. When at last dusk approached, the colonel decided that our platoon-sized headquarters unit would spend the night in one of the wealthy Vietnamese farmers' compounds.

Walking through an arch and into the farmer's courtyard, I was surprised by what I found. A modest flagpole in the courtyard flew the yellow and green Republic of South Vietnam flag. In sharp contrast to the flimsy dwellings I had previously encountered in South Vietnam, here were five large Vietnamese-style houses arranged next to each other in an L shape. Each single-story cement building had a tile roof that extended over a deep shaded veranda, which ran across its entire front façade. The verandas butted up against each other, but were not open to one another.

As the colonel and a few officers entered the middle dwelling to speak

8. Dinner with a Wealthy Vietnamese Farmer

A few 2nd/60th Headquarters soldiers relaxing on a wealthy Vietnamese farmer's veranda after a hot day of chasing the VC.

with the landowner, the rest of us flopped down on the veranda's front steps or leaned on its front wall, happy to have an opportunity to rest. An old Vietnamese woman and young children gradually appeared on the other porches to quietly examine the new arrivals. Again in contrast to my previous experiences, these children did not run up to us and beg. Like the grandmotherly woman, they simply stood on their respective porches and stared at us.

After about half an hour the commanding officers returned to our little encampment. The battalion's FO captain, whose radio I had been carrying, told me what they had learned. The Vietnamese landowner claimed that he was pleased to welcome the Americans to his farm. He reported that the Viet Cong had invaded his farm the day before, demanding many bags of the his rice as a type of taxation. Because he demonstrated some reluctance to comply with their demands, the VC had violently seized his wife and thrown her down the well located on the edge of the courtyard. Of course, the farmer then rushed to pay the extortion to prevent his screaming wife from drowning. The VC then allowed him to lower a rope to his wife and pull her out of the well to safety.

Like most of his fellow wealthy countrymen, this farmer was caught in the middle between the Republic of South Vietnam's aristocratic government and the North Vietnamese communists. Both groups demanded heavy taxation and support from wealthy men. I mentioned his flagpole flying the South Vietnamese flag that day; I am sure that the day before he had been flying the Yellow Star of the communist flag. Basically these wealthy farmers were mainly apolitical, desiring only to be left alone to work their rice paddy fields. Unfortunately, however, his class of wealthy landowners was one of the primary targets of hatred for Ho Chi Minh and his communist followers. It is well documented that upon his rise to power in the North, Ho Chi Minh executed thousands of rich landowners for the supposed crime of being a member of the wealthy upper class.

Another surprise was that our entire headquarters unit had been invited to share an evening dinner with the farmer's family, I think in gratitude for us not committing any atrocities on him or his family. Also, he appreciated the reality that with the American Army in his home, the Viet Cong would dare not return. My reaction was a combination of excitement and curiosity as we washed ourselves and dusted off our uniforms as best we could before the meal. The word was that the farmer was butchering an entire pig for the feast.

Dinner was served in the large room located on the inside of the front door. It appeared to be a general purpose room decorated in one far corner by an altar honoring the family's ancestors. On the altar wall were a few photographic portraits of deceased family members with candles burning before them. The only other furnishing was an immense dining table at least 14 feet long and five feet wide, constructed out of a dark mahogany roughly four inches thick. This majestic table had matching mahogany chairs surrounding it. The Vietnamese family consisted of the farmer, his wife, a grandmother, and two school-aged children, who sat at one end of the vast table. Enlisted soldiers and officers alike occupied the rest of the seats. Curiously, there was an empty place setting and an unoccupied seat at the family's end. Someone asked about it, and our translator informed us all that the chair and plate were for the spirits of their ancestors. Since there was no tablecloth, I was able to feel and admire the heft of the magnificent dark mahogany table.

Our meal consisted of heaps of steaming white rice thoroughly mixed in with the butchered pig that had been completely diced into roughly one-half-inch cubes. Our beverage was simply cold well water served in coarse pottery cups. Before we ate, our host solemnly incanted a Vietnamese blessing to his ancestral gods. Then the smiling farmer's wife served each of us a steaming heap of rice and pork from the large serving bowl in the middle of the table. The food looked and smelled mouthwatering, a far better meal than the heated cans of C-rations I would otherwise have eaten that evening. I attacked my food as best I could with the supplied wooden chopsticks. It was delicious.

Soon, however, I discovered that included in the food mix were diced cubes of bone, pure fat and gristle. The Vietnamese butcher had wasted little of the swine, chopping up the entire animal except for the internal viscera. This discovery was a bit disturbing at first, but none of us complained. I began to recognize the inedible cubes, discreetly moving them to the side of my plate. I fully enjoyed my introduction to exotic Vietnamese cuisine and customs. Judging by the smiles and positive comments of the other men, they did too. This dinner was the only time in my entire yearlong combat tour that I ate with or had any other type of normal social interaction with Vietnamese civilians. American military command did not encourage socialization with the Vietnamese. If command had allowed normal relations, I think there would have been many benefits. Nevertheless, the exotic meal remains a uniquely special event in my life.

That night I was told that the headquarters unit would be spending the night in the landowner's home. I was not told by whom or how this decision was made; however, my theory is that the rice farmer invited us to stay as protection against the Viet Cong returning to torture him or his family. The Vietnamese family soon disappeared, probably moving in with their extended family in one of the other large homes in the compound.

As everyone sought out his own perfect place of protection in the main room or on the porch to sleep, I claimed the mahogany dinner table that had been moved to the side of the room. After inflating my air mattress, I laid it out on top of the table and crawled up on it. I fell right to sleep. Sometime during that night I was awakened by the frightening sound of incoming mortar rounds exploding nearby. The Viet Cong knew we were somewhere in the general area, but did not know our exact location. Consequently, their incoming mortar fire was searching fire that repeatedly shifted around. By that time my combat experience enabled me to judge that the enemy's mortar rounds were hitting randomly and not closing in on us. At that point I simply grabbed my air mattress and rolled down under the thick wooden table. Because I felt secure under the protection of both the thick mahogany table and the house's tile roof, I easily fell back to sleep.

The next morning my captain ordered me to helicopter back to our firebase at Tan Tru so that I could continue with my primary mission of providing safety clearances. While I was on duty at the TOC the following morning, the same captain, who had just returned from the search and destroy mission, informed me of how lucky I was not to have been out in the field the night before. As a realistic precaution the colonel had moved his command group out of the farmer's house and, for whatever reason, into a filthy pigpen. I chuckled internally as I outwardly sympathized with the captain's complaints. We were both glad that the operation had resulted in no friendly casualties.

9

The Helicopter War

America's Vietnam War was the first true helicopter war. Yes, the American Army did use helicopters during the Korean War, but only in small numbers as medical evacuation ambulances for the wounded. In contrast, during the Vietnam War our Army armed helicopters for the first time and utilized them as fighter aircraft. This was the birth of the helicopter gunship.

A few years after the Korean War our military needed a new utility helicopter to replace the obsolete ones still in service. A new era began with the Bell Aircraft Corporation's Iroquois helicopter, which first went into production for the American military in 1960. This modern helicopter was nicknamed the Huey because of the Army's designation of it as Helicopter Utility, One (HU-1). The Huey is a twin-blade, turbine-powered helicopter built mainly of aluminum. Its powerful turbine and light weight made it capable of flying at 135 miles per hour. The Huey lacks wheels; instead it lands on two large skid tubes attached underneath the cabin.

It was initially designed as a medical evacuation and utility aircraft. During the Vietnam War, it was at first utilized as a troop transport helicopter, but early on the Army discovered that while unloading soldiers, the Huey was highiy vulnerable to enemy attacks. The aircraft needed better protection. Consequently Hueys were armed with two 7.62-mm M-60 machine guns at the doors. Over 7,000 of these helicopters were deployed during the Vietnam War. The need for even more enhanced protection also led to Hueys that were specially modified into pure gunships. These models were fitted with rocket launchers and forward firing M-60 machine guns attached to specialized pods. These specially modified Hueys were given the role of air-assault gunships.

The combat effectiveness of these air-assault gunship Hueys in Vietnam led the Bell Corporation in 1966 to design and develop the first specially

Cobra gunship lifting off from our firebase. Notice the tandem seating arrangement of pilot and co-pilot/gunner behind him.

dedicated fighter helicopter, designated as the Attack Helicopter One (AH-1). At first it was nicknamed the snake, and finally the Cobra, because it proved to be so deadly. It was designed to use the Huey's power train and given a characteristic narrow profile by the pilot sitting behind the weapons specialist in a tandem arrangement. In its front turret the Cobra was armed with two 7.62-mm multiple-barreled mini-guns or two 40-mm grenade launchers or a mix of both. The Cobra was also outfitted with rocket pods underneath its stubby wings. Capable of flying 172 miles per hour, the Cobra was spectacularly successful in its roles as an escort and ground attack gunship in the Vietnam War.

My first flight on a Huey was a combat insertion mission with three companies of infantry from our 2/60th Battalion. Early one morning during the dry season, a dozen Hueys were sitting in a line on an empty field at the edge of our firebase. We boarded our Huey lift company's ships, know as the Greyhounds. Once loaded, the helicopters took off one at a time by elevating the tail and then shooting the cabin straight up into the air. It was an impressive sight, seeing the entire flight of Hueys use this maneuver to become airborne in just a couple of minutes.

The cargo doors on combat Hueys were usually removed. I sat on the

Front view of a Cobra gunship, displaying its narrow width that made it a difficult target when attacking head on.

metal floor at the opening for the door with my feet resting on the helicopter's skids. This position provided me with the perfect vantage point to observe the passing landscape. We flew low and fast, over 100 miles per hour, just clearing the jungle's tree tops. In fact, at times our pilot was forced to fly around especially tall trees. What an exhilarating experience, feeling the refreshing wind and watching the jungle whizzing by underneath my feet. This type of low-level flying ensured that we would surprise the VC, who could not see us coming until it was too late. When we arrived on target, we had to jump from the Huey onto a rice paddy as the Huey hovered a few feet above it. Thankfully it was a "cold" landing zone and we received no enemy fire while disembarking.

Our three infantry companies spent that whole day searching the rice paddies for the VC enemy without success. There were no firefights and no causalities. Neither the infantry—affectionately nicknamed the "grunts"—nor I were upset over this fact. When dusk finally arrived, the colonel conceded to the reality and ordered his battalion to return to our firebase.

The Greyhounds returned and we boarded. My flight back was just as exciting as my morning flight out. Again we flew low and fast over the jungle. In 20 minutes we reached our firebase; however, for whatever reason, perhaps because they were as pleased with the day's outcome as I was, the pilots decided to have some fun with the grunts. Instead of simply setting down in the open field, the pilots made their Huey's tail rotors spin wildly at high speeds around and around their cabins, which also spun dizzily before the Huey finally touched down. This wild maneuver caused the infantrymen to suffer whirling vertigo. It was funny watching the grunts jump off of the Hueys and then stagger clumsily, a few falling, as they attempted to walk back to their bunkers.

Hitchhiking was a popular method for American soldiers to travel along the roads of South Vietnam during the war. I frequently hitchhiked on the highways and dirt roads of the Mekong Delta during my combat tour. What may be surprising is the fact that I once hitchhiked a ride on a helicopter. The helicopter was the Hughes Tool Company's OH-6 Cayuse, better known as the Loach. This helicopter's nickname comes from the pronunciation of the acronym of its class of aircraft, a Light Observation Helicopter (LOH).

The Loach was a single-turbine-powered, four-bladed helicopter. It was characterized by its egg-shaped cabin and wraparound glass bubble front canopy. It was capable of carrying two or three men comfortably; however, they had to be strapped in because Loaches had no doors. It was highly maneuverable and capable of flying quietly over 100 miles per hour.

9. The Helicopter War

LOH helicopter undergoing routine maintenance.

In Vietnam, Loaches operated in a variety of roles, for example scouting and transport. In its combat mode the Loach often operated in conjunction with Cobra gunships in what were known as hunter-killer teams.

In the Mekong Delta there were many American firebases and outposts scattered all over the 9th Division's area of operation. One of the methods that commanders used to keep in touch with their scattered units was to have a small number of Loaches transport mail, documents and medicines among the bases. One of these transport Loaches would routinely deliver mail and documents to our firebase from division headquarters in Dong Tam. It would land on a designated heliport near the TOC, usually with only the pilot onboard.

One morning I had to travel to Dong Tam for an administrative matter, so I hung around the heliport at the time that the mail Loach would usually return to division. When the pilot returned to his helicopter, I asked the young warrant officer pilot if I could fly with him to Dong Tam. A warrant officer is a specialized Army rank somewhere between an enlisted man and a regular officer. Warrant officers are only appointed in technical fields like aviation and intelligence. He granted me permission, so I hopped aboard his Loach into the copilot's seat. The glass bubble canopy surrounded me in the front. Since there were no doors, I had to strap myself in with a crisscross safety belt harness. The pilot had me wear the copilot earphones, which enabled me to communicate with him and to listen to the Loach's FM radio.

After a short turbine warm-up we climbed into the air. My young pilot had fiery red hair and a handsome baby face; he looked no older than 19. He was an exciting pilot; he flew his Loach at times over 100 miles per hour, and when he maneuvered, he made sharp turns in which I looked straight out over my right shoulder and found myself looking straight down onto the ground. Those turns made me appreciate being strapped in so tightly. As we continued our flight, the pilot tuned his radio to rock and roll music being broadcast over Armed Forces Radio out of Saigon. In a short while he was swaying our Loach back and forth in synch with the rhythms of the music. The thought crossed my mind that he might have been stoned. I smiled at him, sat back and enjoyed the ride. We continued rhythmically swinging through the air to rock music for about 15 minutes. Those minutes were the most enjoyable I ever had riding a helicopter in Vietnam. We were dancing in the air.

Suddenly red tracers blazed down directly in front of our canopy. I spotted a Cobra gunship at altitude firing its miniguns as it started its attack run on a VC ground target. We had inadvertently flown into the war. My pilot spotted it too and immediately executed a severely sharp turn, shouting into his microphone, "We need to get out of here." And we did, at speed. The pilot turned off the music. We flew the rest of the way to Dong Tam silently as we both realized how close we had come to disaster.

Near the end of my combat tour in January 1970, I was given an opportunity to fly again in a Loach as a reward for honorable service. A friendly lieutenant from my hometown arranged that I and another short timer were taken on an observation flight over a number of towns in the Mekong Delta. I was terrifically excited about the flight for a number of reasons, not the least of which was that I had recently bought a used 35-millimeter camera. We boarded our Loach at one of our firebase's helipads one bright sunny morning. Our pilot informed us that our flight was officially recorded as a reconnaissance flight.

He flew his small helicopter low and slow over the town of My Tho and its surrounding waterways. We flew a bit above rooftop level over some other small delta towns. It was interesting to watch Vietnamese civilians carrying out their routine daily chores of cooking or washing clothes in the river water. They were so used to helicopters flying over that few bothered to look up. Many of their small houses were built out of tin panels

9. The Helicopter War

and were what we Americans would classify as shacks. Many of the Vietnamese lived an amphibious existence on houseboats moored to the riverbanks, while others had boats that were actively used for fishing. Full of excitement, I snapped an entire roll of film. Our flight lasted about 45 minutes and then the pilot returned us to our firebase. I thanked our pilot exuberantly. Sadly, none of the photographs I tried to take on that Loach flight came out. My new camera was defective, a major disappointment.

The major liability of the Army's many Vietnam War Hueys was that they were highly vulnerable to enemy ground fire. Helicopters are not stable flight platforms; like bumblebees, they are not supposed to fly. Unlike an airplane that may glide to a landing if the motor quits, a helicopter without a functioning rotor blade falls like a brick. The communists shot down our Hueys with machine guns, rocket propelled grenades, AK-47 rifles, and even pistols. All that a Viet Cong fighter had to do was damage the tail rotor of a Huey with a pistol round and the helicopter would subsequently spin itself into a crash, frequently killing everyone on board. It is reported that the American military lost 3,000 Hueys in the Vietnam War, 43 percent of the 7,000 deployed.

Our colonel's Huey nearly had a disaster on New Year's Eve 19970. At the stroke of midnight, nearly every trooper stationed at our firebase shot a hand held flare into the air to celebrate. The flares shot up about 150 yards and then parachuted down slowly, burning brightly the entire time. The different colored flares were our version of holiday fireworks. The colonel's Huey was landing at that time and one of the falling flare parachutes got tangled in his Huey's tail rotor. Fortunately, his Huey was near the ground when the entanglement occurred and the pilot was able to control the helicopter for the few seconds before it landed heavy and hard. I was told that the colonel was furious over the near disaster. He ordered his sergeant major to find and punish whoever had shot the flare. The sergeant major failed because the next morning he could not tell who had launched that particular flare among the hundreds and hundreds of parachute flares lying all over the grounds of our firebase.

The only time generals from our division were killed during my Vietnam tour was when they were flying in Hueys. Whenever any major ground battle broke out in our region, the air overhead would become filled with high-ranking officers flying in their C&C Hueys, trying to direct the battle from above. A poor method of command, I thought. One of our sister infantry battalions, which was on a joint operation with the South Vietnamese Army during Nixon's Vietnamization program, found and engaged a large VC force. This new type of combined forces battle brought out not

Huey helicopter resupplying one of our 2nd/60th Infantry companies in the field.

only the American and the Vietnamese battalion commanders' C&C Hueys, but also American and Vietnamese division generals flying in their C&C Hueys. All these command helicopters were circling in the tight air space over the battle zone, which was crowded with other Hueys ferrying in reinforcements. There were also Cobra gunships engaged in the battle.

In the resulting air space chaos, two C&C Hueys crashed into each other. The commanding generals and their crews were all killed. I was told that Huey parts fell from the air like bricks. This tragedy significantly disrupted the command structure of our division, as colonels were quickly promoted to generals and majors were promoted to fill the empty colonel slots. Of course, the combat effectiveness of our division fell as a result. This tragedy also reveals another inherent weakness of our military structure in Vietnam. Our Army was top heavy, with too many layers of command. Those division generals should not have been interfering with their battalion commanders. It was as if there was not enough war to go around. We had too many chiefs and not enough Indians.

As mentioned, Vietnam War Hueys were equipped with two M-60 light machine guns attached to the aircraft at the openings of the cargo doors. Two door gunners had to be added to the crews to fire the machine guns. Army regulations stated that only infantrymen were eligible to volunteer to be a Huey door gunner. This proved to be an endless topic of controversy and conversation for the average infantryman.

Every infantryman I talked with had his opinion about volunteering for the job. Some considered the door gunner's job to be much more dangerous than that of an infantryman. They would argue that the door gunner was highly vulnerable because he was out in the open. Door gunners had nothing to hide behind and could not dig a foxhole for defense. Also, these infantrymen knew that the Hueys and their door gunners were highly valued targets for the VC and NVA. On the other hand, many infantrymen did consider volunteering for the job because a door gunner did not have to deal with the mud and dangers of ground combat. These men were jealous of the door gunner's ability to sleep in a bed every night rather than on the ground or in the mud. They would argue that door gunners did hide behind the armor plated vests that many gunners were issued. And finally there was the glamour of flying in the sexy Hueys. After listening to their ongoing arguments, I concluded that it was a difficult choice that I was glad I did not have to make. If forced, I would probably have opted to volunteer to be a door gunner. Flying in a Huey was that exhilarating to me, plus I hated sleeping in the mud.

10

Rocket Attack

During my combat tour in South Vietnam, I could have been wounded or killed by enemy fire many different times. However there were two distinct incidents when I came the closest to Death's embrace.

The first event occurred during the wet season on August 12. Because I had been on duty the entire night before, I slept in during that day. My bed was the bottom bunk of an Army-issue olive green metal bunk bed. I had been issued a thin mattress that covered a steel spring frame and regular white sheets that I treasured. As usual, the temperature was in the high nineties and sticking humid, but I had recently bought a small fan at the Saigon PX that provided a little cooling comfort.

After awakening in the late afternoon, I hung around our bunker socializing with my bunker mates, waiting for dinnertime. I was sitting on my wooden footlocker next to my bunk, when I first heard a distant muffled thump. Someone yelled out, "Incoming," which meant that the explosion was not our own "outgoing" artillery fire at the enemy. The Viet Cong were firing rounds at us. At once I went into an emergency alarm state, my body tensed and all my senses more acute. At that point my primary emotion was an alarmed alertness.

Then I heard another, louder incoming explosion. My body's tension increased when I recognized that this new explosion was closer than the first one. I estimated that the enemy rounds were hitting "outside the wire," which meant that they were not falling inside our firebase's perimeter fortifications. My total consciousness centered on straining to hear the next rounds and explosions that I knew were coming. The Viet Cong never fired just one or two rounds in an attack, because of the risk of detection they took by attacking us. If they were going to risk an attack, it would be with multiple rounds. I noticed that all the men in my bunker had stopped what they were doing and, like me, were listening intensely. Those that had been

10. Rocket Attack

standing sat down near their bunks. Men who had been listening to music turned their radios off. All the usual chatter stopped. Since we had no partitions between our bunk beds, I could see the alarmed anxiety on all of their faces.

WUMP! BOOM! The third round hit and exploded. Now I realized that this attack was a serious one. This latest explosion was near enough that I heard the low sonic boom caused by the blast's rapidly expanding gases. I estimated that this round had hit near or on our barbed wire perimeter defenses. That reality was bad enough, but what was more upsetting was that this new explosion was aligned with the previous two. The Viet Cong shooting at us were walking the rounds toward us in a straight line. Mortars are difficult to aim because of their smooth barrel and high angle of trajectory. Under mortar fire the rounds fall in a rather random pattern over the general target area. This aimed straight-line pattern, combined with the power of the explosions, demonstrated to me that the enemy was not firing mortars, but rather the more deadly rockets.

The Chinese communists had been supplying their Vietnamese allies with thousands of their 107-millimeter rockets. Consider the fact that the standard size of the U.S. Army's most common howitzer is 105 millimeters, which is the diameter of the gun's barrel. The 107-mm rocket was comparable to our 105-mm howitzers in terms of velocity and effect. Our 105-mm howitzers are capable of launching a 32 pound shell to the maximum range of seven miles. The Chinese 107-mm rocket weighs 42 pounds, is electrically triggered, and carries a fragmentation warhead. Its maximum range is five miles. The Viet Cong were now attacking my firebase with artillery-sized weapons.

The Chinese 107-mm rocket could be fired from a multiple tube or a single tube launcher if available; however, the Viet Cong usually built an improvised, field expedient one. The most common launcher was simply two crossed sticks, which were stuck into the ground and then lashed together with vines to form an asymmetrical X. The upper section of the rocket would rest on the lashing at the crossing point; the tail end of the rocket would simply rest on the ground. Although the Viet Cong sacrificed accuracy with the use of a field expedient launcher, this use of an improvised one provided them with a number of advantages. The materials used to construct the launchers were readily available and could be assembled easily. However the most important advantage was that the VC could launch their rockets rapidly and then immediately withdraw. They were not delayed by retrieving valuable equipment; consequently, they were usually able to avoid detection and our subsequent counter-rocket fire.

Artillery-sized rockets were being fired at us—at me. These rounds were significantly more lethal than the VC's usual mortar attacks. This attack became deadly serious. The alert tension I had been feeling transformed into outright fear. I was sitting on the floor of our bunker with my back to the wall when BAM! A fourth rocket exploded. This one had definitely hit within our perimeter. It too was aligned with the previous explosions, forming a straight line aimed directly at our bunker and me. My heart pounded rapidly and my mouth dried out. I found my flak jacket and put it on. These Vietnam-era flak jackets looked like, but were not, bulletproof vests. Heavy and bulky to wear, they were designed to offer protection against flying shrapnel. Then I noticed that most of the men in our bunker had donned their flak jackets too. It was eerily quiet; no one was talking. Fear was visible on their faces.

I was thinking that I was nothing more than a target to be shot at, like a tin duck in a carnival's shooting gallery. There was no fighting back; there was no way of stopping the incoming rockets. I wished that I had been on duty that day at the TOC. If I were on duty, I would be active and distracted. I would have been part of the team trying to determine where the rockets were coming from. If I were on duty, I would have been involved in the launching of our own counter-rocket fire. I would have been fighting back. But no, I was just sitting there, a helpless target, growing more and more frightened with every new rocket explosion.

It became clear to me why the psychological disorder called "shell shock" was first associated with artillery barrages in World War I. Under artillery or rocket fire, a soldier is a vulnerable human target who can only shelter as best he can. There is no fighting back. It does not matter how strong or tough the soldier is; it does not matter how well trained he is; it does not matter that is he is a member of a prestigious unit like the Green Berets. When a large rocket is screaming down towards his head, being a rough, tough elite trooper means nothing. He is just another vulnerable human target, another tin duck down range at the shooting gallery. Army recruiters and recruitment advertisements never mention this part of the military experience. No one is invulnerable to artillery or rocket fire.

BAM! The fifth rocket exploded. The blast was thunderous. Horrifyingly, it too was on the straight line with the previous ones, forming the deadly arrow pointing directly at me. Now I had no doubt that the next rocket would hit our bunker. For the first time in my life I felt pure terror. I could actually be killed in the next few minutes. My anus tightened involuntarily. I felt hollow inside, like the bottom had fallen out of my stomach. This could be the moment of my death. I had to do something. I roll under

10. Rocket Attack

Damage to support beam of living quarters bunker from a communist 107-mm rocket that exploded eight yards from the bunker I was in.

my bunk, which had a steel frame and a mattress. I hoped the added layers of protection would make the difference in saving my life. I stopped breathing.

BAM! The sixth rocket hit. The explosion was ear shatteringly close. I could hear the red hot shrapnel pounding the side and roof of our bunker right at the point where I was lying. Thankfully, none of the shrapnel penetrated the dirt-filled wooden ammo boxes stacked against the sides of our bunker. Lying under my bunk, I was breathing easier, yet I still feared another incoming rocket. But none came. The Viet Cong had fired only six rockets. I was sure that they were now fleeing. I heard our artillery and mortars shooting counter-rocket fire. I hoped it would kill the Viet Cong, but deep down I knew they were gone. Their pattern of attack was shoot and scoot.

My bunker mates and I were curious as to where exactly the last rocket hit. After a few minutes of quiet, we retrieved our nerve and went outside to explore. We discovered that the last rocket had hit the next bunker, which was ten yards away across the dirt road that separated it from ours. The soldiers living in that bunker were fortunate that the 107-mm rocket was at the end of its effective range. Its trajectory was arcing downward when it hit one of the one-yard by one-yard square pillars that supported the bunker. Because we were in the Mekong Delta with its shallow water table, the Army's combat engineers could not dig down into the earth to build bunkers. Instead they had driven large wooden support pillars into the mud at each corner of the bunker as well as a few in the middle. The last Viet Cong 107-mm rocket struck a corner support pillar, splintering it. If that rocket had flown straight into the side of the bunker at full velocity, it could have penetrated the ammo box barrier and exploded inside. Many of my 9th Infantry comrades would have been killed or wounded. Both they and I were lucky that terribly frightening day in South Vietnam.

11

Losing the War

During the end of the 1969 wet season, two of our battalion's infantry companies were pursuing at least a company of Viet Cong fighters. The enemy had taken a number of casualties and was retreating into the Plain of Reeds region that encompasses the border between the countries of South Vietnam and Cambodia. The area is a desolate lowland swamp covered by tall grass reeds and flooded by small waterways that crisscross the entire landscape.

Throughout the battle, our colonel had ordered continued reinforcements of his infantry companies. Because the battleground was located far north of our usual operational area, he had also ordered that two 105-mm howitzers, their crews, the FDC personnel, and a large supply of 105-mm artillery rounds be airlifted to the battleground area. These guns would then be able to provide the umbrella of artillery protection that our infantry soldiers could always rely upon. Protective infantry support from airplanes is not effective at night or when the weather is stormy. In contrast, artillery, when in range, can always supply heavy firepower protection to any endangered infantry.

Near the end of the day I was ordered to resupply our battalion's FO with the critical radio batteries and to function as his RTO as part of the overall reinforcements. I was airlifted into a cold landing zone. The ground fighting had eased to a halt when the enemy unit retreated all the way into Cambodia, a no-go zone for our military. To this day, even after much study, I do not understand why America did not pursue the enemy into their Cambodian sanctuaries. Cambodia was not the neutral country the American media portrayed it to be during the Vietnam War. The North Vietnamese were shipping supplies to Southern Cambodian ports, where they were then transported by small boats and by foot to Vietnam. The infamous Ho Chi Minh supply trail ran down through Cambodia. Later, when President Nixon did order a limited excursion into Cambodia, our nation's campuses erupted in protest, but those student protestors did not fully understand the realities of the Vietnam War.

Once on the Plain of Reeds, I was stationed at the battalion's headquarters base area, not with the howitzers, which were more protected by their location in our rear area. My FO told me that we would all spend the night in the field and then be airlifted back to our Tan Tru firebase the next morning. However, he reported that we did not have enough of the always valuable "blade time" for all of the 105-mm rounds to be airlifted back as well. In other words, we did not have enough helicopters to bring back all the artillery ammo, and besides it was always dangerous to transport artillery rounds by helicopter. Instead he ordered me to fire off the artillery ammunition we had that night.

In response to these orders, I contacted our FDC crew and we began the process of plotting targets of suspected enemy positions in Cambodia. I knew there was a large enemy base across the border because the previous week one of our FO captains had flown a reconnaissance mission over the Plain of Reeds. He told me he had circled the area wide and inadvertently crossed over the border. As soon as he did, he started to have enemy antiaircraft guns firing at his plane. Those guns had to be protecting a base camp.

A few minutes later, when my FO captain discovered that I was planning for targets in Cambodia, he lost emotional control. My immediate superior was a West Point graduate who was extremely ambitious. He always considered any tactical decision in terms of how it would affect his military career. Screaming at me, he ordered me to stop plotting those targets and to fire the excess rounds to the South. Not believing what I was hearing, I protested, "But sir, we are in the middle of nowhere. Who is going to know?"

He refused to budge: "We would be causing an international incident. I cannot let that happen." I was sure that his decision was all about not jeopardizing his future military career. How could he ever be promoted to general, if there were an international incident blot on his record?

Still yelling, he reordered me to fire the 105-mm artillery rounds to the South of our position. Like a good soldier, I answered, "Yes, sir." Contacting FDC again, I informed them of our new target direction and we plotted the new targets to the South. The appropriate Viet Cong targets were to the West, yet we fired over a hundred expensive 105-mm howitzer rounds to our South. We killed a lot of nipa palm trees in South Vietnam that night.

After the FO captain finished screaming at me, I came to the unsettling conclusion that we would not win the war in Vietnam. I thought, "How can we win this war, when an elite officer orders me to fire one of our most powerful and lethal weapons, artillery, in one direction, when he knows that the enemy is hiding in another direction?" It simply did not make common sense. His decision made me wonder again, "What kind of Alice in Wonderland, rabbit-hole craziness is this war in Vietnam?"

12

Time on Target Attack

Artillery is called the "king of the battlefield" for a good reason. Most people, if asked the question of what was the most lethal American weapon in World War II, would probably answer "the atomic bomb" because of its fearsome power. They would be wrong. Post-war casualty counts revealed that that the firebombing of Japanese cities by our B-29 bombers caused more enemy deaths than the atomic bomb. Yet there was an even more deadly weapon. Our instructors in AIT taught us that during World War II, more soldiers were killed by artillery fire than by any other weapon.

Just one of our 105-mm howitzers could deliver devastating lethal destruction on the Viet Cong. Our one battery firing all six 105-mm howitzers at an enemy position could produce significantly more destructive power. However, the most overwhelming destructive firepower I was involved with during my tour in South Vietnam was in an artillery mission known as a "Time on Target" attack.

One night my artillery commander, the battalion FO captain, who was another West Point graduate, called me over to the large area of operations map that was on the wall of the infantry section of our TOC. He pointed to a location on the map and informed me that recent intelligence reports had pinpointed it as a likely enemy position. He ordered me to organize a TOT attack on this target for the next night. It was the first time I had even heard of this type of artillery fire mission, so I admitted my ignorance. He explained that I was to organize and coordinate a multiple battery fire mission designed to ensure that all the rounds fired from the different batteries would strike the target at exactly the same instant. This synchronization of multiple battery fire was called a Time on Target (TOT) attack. Listening to the captain, I quickly realized that this type of fire mission was an extremely complex one that would require split second timing.

12. Time on Target Attack

As the captain continued explaining the TOT mission, it became obvious to me that this special mission must have been ordered at the highest level of military authority. First of all, I was to coordinate six different artillery batteries from five different firebases. This meant that 36 guns would be involved, the most howitzers I had ever fired at any one time. Secondly, I was given authority to utilize the battery of 8-inch howitzers firing out of the Division's main base at Dong Tam, as well as another battery of 175-mm howitzers. This mission was the first and only time I had access to these extremely potent artillery weapons. An 8-inch howitzer's shell weighs 200 pounds. The gun can fire rounds out to a maximum range of ten miles. The 175-mm howitzers were called Long Toms because of their elongated barrels. Their shells weigh 174 pounds. The Long Tom can fire rounds to a maximum range of 20 miles.

The TOT's real complexity comes not from firing multiple batteries at one target, but rather from coordinating the fire so that the rounds all strike the target at the same time. Since the six batteries assigned to this mission were located at different firebases, their distances from the designated target were different. This discrepancy meant that the flight times of the fired projectiles would be different. A synchronized instantaneous strike required that the more distant howitzers fire first. Only then would the next closest guns fire, and so on with the closer batteries.

My first step in executing the captain's orders was to contact the various batteries through landline telephones for increased security versus FM radio transmissions, which the enemy could monitor. Once connected to the FDC centers of the batteries involved, I gave them the target's location so that they could then compute the trigonometry required for the firing solutions. Once the solution was found, they were to report to me their rounds-to-the-target flight times. When I had acquired the times from all six howitzer batteries, I then calculated which battery would have to fire first, which one would fire second and so on to ensure a simultaneous strike.

On the night of the TOT attack I experienced one of the most memorable and dramatic moments of my entire Vietnam combat tour. Because I was ordered to coordinate the attack, I had the responsibility of giving the firing orders to all of the howitzer batteries. At the designated time, with the radio microphone in one hand and a stop-watch in the other, I called out the firing orders in the exact sequence of time intervals I had memorized from my calculation: "Dong Tam fire! .... Rach Kein fire! .... Ben Luc fire! .... Tan Tru fire! ..... Binh Phtoc fire!"

As I finished ordering the firing sequence, I could not help but imagine the rounds flying through the air and the apocalyptic carnage of the actual strike. Thirty-six massive artillery warheads simultaneously exploding would cause a concussive shock wave blasting out hundreds of yards. Even enemy soldiers hiding in bunkers would have their lungs implode, then explode. Thousands of red-hot steel shrapnel fragments would explode out at over 1,000 feet per second, slicing and skewering everything in their path. Any unprotected Viet Cong or North Vietnamese Army soldiers within hundreds of yards would be torn to pieces, if the shock wave did not kill them first.

For one half of a split second I felt a sense of pity for any human being caught under that artillery deluge. The butchery would be ghastly. It was the first and only time in the Vietnam War that I felt any such sympathetic feelings regarding the enemy. But the emotion instantly faded because my very next thought was that possibly the intelligence information was wrong, as it usually was. There were probably no communist soldiers at the receiving end of that huge TOT artillery strike.

I never received any after-action report of enemy casualties caused by that TOT attack, which argues that no one ever found any. Or perhaps there never was a valid casualty count in the first place, because all that remained at the target would be a center of charred and cratered earth surrounded by the splintered and toppled trees.

13

Firebase Animals

Rats

American soldiers in Vietnam not only lived like animals, but also lived with them. Native to South Vietnam were beasts such as tigers, elephants and small gorillas that the Marines nicknamed "rock apes" because at times the primates would throw rocks at them. My animal encounters were not with these exotic species; instead I was forced to live with rats, snakes, monkeys, dogs and water buffalo. Also we had to cope with the swarms of mosquitoes that thrived in Vietnam's Mekong Delta.

The rats were the worst. The Mekong Delta's numerous rice paddies sustained an enormous population of large wild rats that fed on the rice shoots. Once American soldiers arrived, our food scraps and garbage became accessible, resulting in an invasion of these ugly, disease-ridden rodents into our mess halls, tents, bunkers and garbage dumps. When I first arrived in country, the reality of having to live among these vermin frightened me because visions of the plague ran through my mind. Later I learned that the real threat was rabies after a rat bite. At the time—1969—any soldier whom an infected rat had bitten would have to endure the standard medical treatment of the era. The treatment consisted of a series of daily injections of the rabies antidote into the soldier's abdomen. I understood that these injections were quite painful, and I hoped to avoid the treatment at all costs.

Because the Mekong Delta is a low-lying swamp, our combat engineers built our living quarters bunkers on large wooden pillars that were driven deep into the mud. Thick wooden planks were screwed on top of the pillars to form the bunker's floor. These floorboards were laid out with sizable gaps between the planks, as a carpenter would do building an outdoor deck. Once the bunkers were inhabited, our food scraps and other edible debris fell into these gaps to land in the mud below the bunker's crawl space. These scraps attracted hordes of rats that feasted on them.

A jeep stuck in the mud of the dirt road outside our living quarters bunker during the rainy season.

Many of the bolder rats would crawl up onto the wooden beams that formed our bunker's roof. Some rats even made it into our living areas. My sniper roommate reported being shocked awake one day by a rat that ran across his hand. My favorite FO captain told me that he awoke one morning to find a rat sitting on his chest staring at him. Of course he freaked out, jumping up to try to kill the rat.

My worst experience with these repulsive rodents occurred on my July 7 birthday. My fiancée was kind enough to send me her homemade birthday cake. After traveling halfway around the world, the cake was delivered on July 6 inside a thick cardboard cake box with an attached birthday card. Since my birthday was not till the next day, I decided not to open the box.

Instead I stored the box on a wooden shelf above my bunk. I was looking forward to cutting the cake and sharing it with my bunker mates.

But the next day when I took down my cake box and opened it, I was horrified to discover that rats had eaten through the bottom corner of the box in order to devour the bottom one-quarter of the cake. The top of my birthday cake was still intact. My next decision demonstrates how the personal values of a soldier change because of war. Most civilians would have thrown the entire cake away in disgust. My reaction was, "There is still plenty of good cake left. One half of the cake for the rats, one half for me," and I took my bayonet and cut the cake in half. A few friends and I all enjoyed that special twenty-third birthday cake half in my small bunker room. Every one of us appreciated this rare treat of homemade cake.

The best way to control the rats was to have a dog live in the bunker. One of my bunker mates adopted a small, totally brown terrier puppy. Obviously he named her Brownie. She became the bunker's pet dog; we all took care of her. With unending energy and enthusiasm, Brownie accomplished what her breed was bred for: she loved to chase and kill the rats. In a short time the rats learned that they had to avoid our bunker's living area to avoid Brownie's bite. After she moved in with us I never saw another rat in our bunker. However, they could still forage in the crawl space under our floor. When they did, Brownie would go wild, running in all types of zigzag patterns, barking all the time as she followed the scent of the rats below her. As I had painfully learned, they also crawled up onto our roof beams where Brownie could not follow. However, no one in our bunker had to suffer the pain of a rat bite while Brownie lived with us. I had real affection for little Brownie.

There is an ironic epilogue to my Vietnam rat experiences. During my first year back in the States after Vietnam, I began graduate studies in clinical psychology at Indiana University. My assistantship required that I assist a psychology professor who taught undergraduate experimental psychology classes. Unbeknownst to me when I accepted the position, I was required to "gentle" the white laboratory rats used in the class's experiments. Before students were allowed to work with these white rats, they had to have human hands hold and stroke them in order that the lab rats would become used to human handling.

At first I was reluctant to perform the gentling, especially after I was told that I would not be allowed to wear thick gloves. The handling had to be performed by bare hands. But the professor reassured me that this breed of white laboratory rat was bred for gentleness.

Consequently, I summoned my courage and went to the animal housing room in the psychology building the next Saturday morning. I picked up the first white rat and stroked him a few times before returning him to his cage without any problems. Perhaps my fears were overblown. I picked up the second rat and he immediately bit me on my index finger. I screamed in shock and pain as the rat hung from my right hand with its teeth embedded in my finger. The pain and the spectacle of the rat hanging from my hand enraged me. I grabbed the rat with my left hand and threw him as hard as I could against the closest tile wall. The rodent slammed against the wall and slowly slid to the floor stunned. Then I jumped up and landed with both feet on the offending rodent's back, killing him. That rat would never bite another human being.

Still shuddering with a fear of rabies, I quickly disposed of the rat's body by dumping it into the special container the psychology department had for dead experimental animals. Then I ran to the student health clinic a block away, with the specter of rabies and its painful treatment dominating my consciousness. Once I arrived, the nurse calmed me down by assuring me that these white laboratory rats were rabies free. She gave me an injection to prevent tetanus and then we had the following conversation:

> NURSE: "There is a state law in Indiana that after any rat bite the animal has to be quarantined for 48 hours for observation. Bring the rat back here to our clinic."
> ME: "The rat is dead."
> NURSE: "What happened?"
> ME: "I killed it."
> NURSE: "You brute! Well now the brain of the rat will have to be sent to the Indianapolis lab for a rabies assessment. Get the body and bring it back here."

I retrieved the rat's body from the dead animal container and brought it back to the student health clinic as directed. The same nurse brought out a little guillotine device for severing the rat's head, but she was squeamish about the procedure. Once I recognized her reluctance, I immediately volunteered to do the job. Carefully I placed the rat in the device with its head under the blade. Then I punched down on the guillotine to cleanly chop off the head of the rat that bit me. The paradox of my killing that rat twice provided me with some Vietnam War style payback satisfaction against the entire rat species for those that ate my birthday cake.

By January 1970 I was in the last stages of my yearlong Vietnam combat tour; I was a "short timer." One morning after my nightlong duty at the TOC, I went to the grunt mess hall for a late breakfast. The grunt mess hall was a giant magnet for the hordes of rats that lived in our firebase. The smell of the food and the scraps in the nearby garbage dump attracted them by the dozens. A popular form of entertainment for soldiers at our camp was to go to the garbage dump at night with flashlights and pistols to shoot the rats. One California hippie type guy in our bunker used to hunt garbage dump rats with a bow and arrows. I never participated; I never was much of a hunter.

At that particular time in the morning, the mess hall was basically empty because the serving time for breakfast was just about over. I filled my plate with a serving of scrambled powder eggs and retreated to a back table to eat alone. As I was eating I spotted three large black rats on top of a nearby table eating scraps of food. I did not move; I was too tired to jump up and attack them. I stared at them; they ignored me. I decided that if they stayed on their table, I would stay at mine. As I finished my eggs I told myself, "You have been here entirely too long. It is time for you to go home."

Dogs

South Vietnam's dogs were the animals that were the most fun to live with. There is a famous photograph published in American newspapers and magazines that featured two South Vietnamese soldiers walking away from the camera, each with a small dog on a leash. Most Americans interpreted the photograph as a cute one, soldiers walking their pets. They would have been disgusted by the truth. Those dogs were not pets; they were dinner for the Vietnamese soldiers. Vietnamese ate dogs; Americans fed them. Guess which side the dogs were on during America's Vietnam War?

A number of soldiers at our firebase adopted dogs during their Vietnam War tour of duty. They were not allowed to take their pets back to the world with them. Most men simply abandoned the dogs when they rotated home. Other soldiers would invariably feed the dogs left behind. Consequently there was a pack of independent dogs that ran free all over our firebase.

During my tour I learned to pay close attention to the warnings these canines provided. Not once during my entire tour did our firebase come under attack after our intelligence officers had warned us that an attack was imminent. The attacks never materialized as predicted. In contrast, after one ground attack, I was reviewing the incident in my mind when I recalled that the free-range dog pack had been barking wildly beforehand. They had picked up the scent of the enemy VC guerrillas. After that lesson I always took the necessary precautions for an attack whenever the dogs started barking wildly. They were our best guards.

These dog packs enjoyed a primitive lifestyle. For a few weeks during the dry season the dogs engaged in free-for-all brawls. There were no soldiers forcing them to fight, like some sick dog owners in the States too often do. The firebase dogs would spontaneously start the brawls. Eight to twelve dogs would all growl, snarl and bark as they attacked each other in a wild noisy melee. One dog might latch on to another's neck and hang on while the other dog tried to shake him free. They would raise a cloud of dust with their high-speed attacks and maneuvers. It appeared that they enjoyed fighting. Once I saw a medium-sized black mutt enter the fray, bouncing in sideways with his back arched like a cat. My theory is that the dog had fought with cats that fought sideways, and consequently he experienced what an effective strategy the side attack was. Dogs would be bitten,

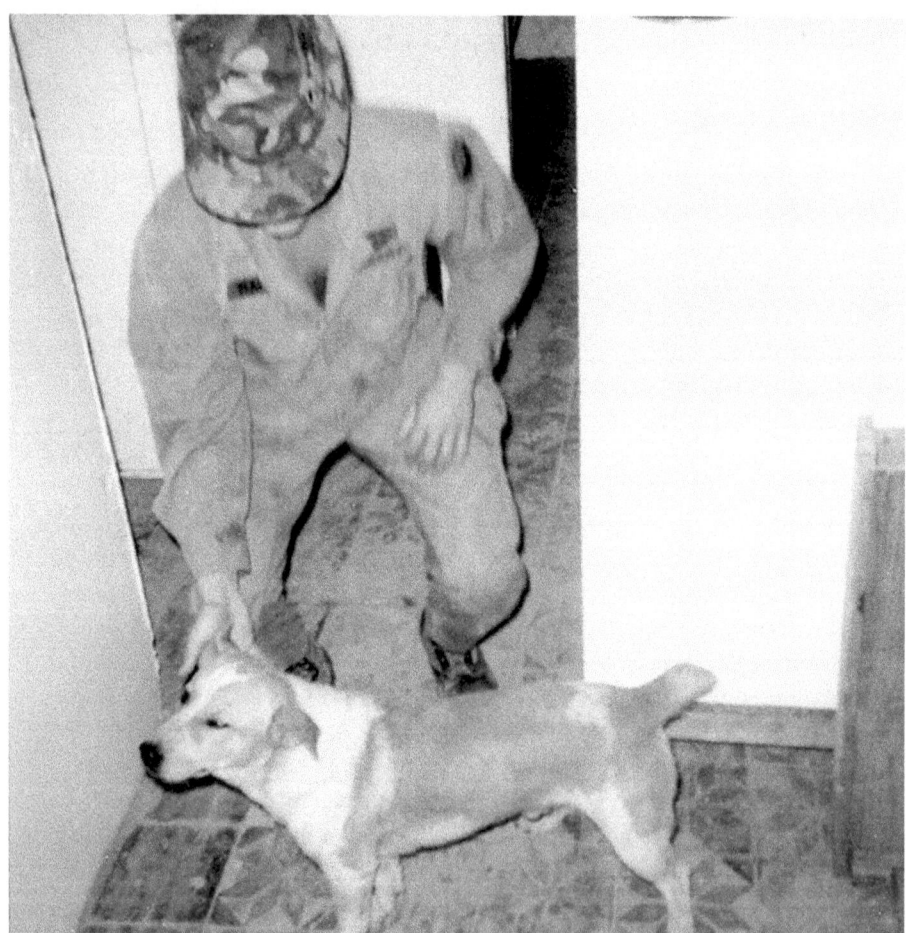

Tu Di, one the toughest dog brawlers.

there would be yelps of pain and sometimes a little blood, but no dog was ever killed or seriously injured. The mutts seemed to somehow view the fights as a type of sport; they never crossed the line into deadly violence. It is a sad comment on humans that we do not follow their example.

One dog that was a regular in the donnybrooks was Tu Di, who was the pet of a disliked and disrespected infantry officer. Tu Di was an odd-looking mongrel that was an eclectic mix of different breeds. He was smaller than the average mutt, with the low-slung body and short legs of a terrier. Yet he possessed the thick chest and big jaw and head of a sizable breed of dog such as a German shepherd. He was so misshapen that when he had an erection, his penis would drag along the ground as he ran after bitches.

Tu Di's life was troubled. Because the enlisted men who hated his officer/owner would often go out of their way to tease or kick him. It made him mean and nasty. He would enter every dogfight by charging in at full speed to crash into a bigger dog with his broad chest. Frequently he would knock over his opponent; then Tu Di would clamp down on his adversary's neck with his huge jaw. Once I saw Tu Di lock his jaw down on a much larger dog's neck. That dog swung Tu Di's entire body back and forth in the air to shake him off. He failed.

The dogfights were loud and exciting, with tons of action. And remember, no dead dogs. We were so deprived of entertainment that these dogfights became a daily sporting event, the happy hour thing to do.

The Monkey

There was, however, another animal that dominated the dogfights. It was a squirrel monkey that lived in the roof of the Military Police bunker next to ours. Like the feral dogs, this small monkey was more of a firebase resident than any one soldier's pet. The monkey loved to drink the beers that men would offer him. He often acted inebriated. For example, one night he suddenly jumped down from his perch onto the shoulders of a sergeant E-6, who was stumbling back to his bunker late after drinking at the NCO club. The tipsy sergeant screamed out that the VC were attacking him, waking many men. They all laughed at him when the truth about the attack was discovered.

Even though he was much smaller than any of the dogs, our monkey, because of his more complex primate brain, was able to dominate the canine free-for-all. He would run into the middle of the fray, smack the biggest dog on its snout and execute a hasty retreat. Howling in pain, that dog and his allies would take off after the monkey, who headed straight for his home bunker. Someone had set two one-half inch planks into spaces between the ammunition boxes stacked along the bunker's wall. These boards were like spring boards for the monkey, who would jump on the lowest one, which then bent like a diving board and flung him up to the highest board in a split second.

From the highest board the monkey would jump up to his nest in the roof rafters.

Feeding the monkey that lived in the roof of the MPs' bunker. Notice the wood planks set in between the ammo boxes that provided a springboard ladder for the monkey.

The dogs chasing him would try to crawl or jump up the ammo box wall in angry desperation; however, dogs are not climbers. All their wild efforts failed. In the meantime, the monkey was driving the dogs even crazier by squawking at them while he threw his own shit down on them. The growling and howling dogs would reach a state of savage delirium, while the monkey hooted in delight. It was always a laugh to watch the dogs and monkey battle.

Mosquitoes

In contrast to dogs, the one insect on my critter list, mosquitoes, trigger nothing but bad memories. In the Delta these disease-ridden predatory insects were large, aggressive and innumerable. The wet swamplands of South Vietnam's Mekong Delta were ideal breeding grounds for mosquitoes; there seemed to be hundreds of thousands of them inhabiting the area around our firebase. One night, while walking to the TOC, I saw thousands and thousands of them swarming around a light at the top of the officer's club hooch. They formed a thick buzzing cloud of disease that no sane creature would want to walk or fly through.

The swarms of mosquitoes brought with them the threat of malaria for those they bit. It was common for American soldiers in the Mekong Delta to become infected with the disease. Victims suffered horribly. The set of miserable symptoms included chills, sweats, fever, fatigue, headaches,

13. Firebase Animals

and vomiting. The Army mounted major campaigns to limit the scourge of malaria. Mosquito repellent was widely available; just about every infantryman in our battalion had his bottle of insect repellent strapped to his helmet. The Army also issued cans of aerosol bug spray to kill the insects. It surprised me to read on the can's label that the spray was an anti-cholinergic agent, which biologically means that it was basically a nerve gas that paralyzed the mosquito's nervous system. It worked wonderfully well; unfortunately it seemed to be toxic to humans too. When I sprayed it all over our little bunker cubicle, I had to leave for about 15 minutes. A human could not breathe in the cloud formed by the bug spray. Upon returning, I would find dead mosquitoes all over our bunk beds and the floor.

The most disliked weapon in the Army's war on malaria was the orange malaria pills. These pills were a little bigger than a quarter and three times thicker. They were prophylactic medications, designed to prevent the soldier from coming down with malaria even if an infected mosquito bit him. The military's plan was to have every American soldier in South Vietnam swallow one of these large pills a week. However, there were problems associated with the pills. Besides being difficult to swallow because of their size, they also caused painful stomach cramps with accompanying diarrhea. These side effects were so uncomfortable that many soldiers refused to ingest the pills. This lack of compliance was so widespread that on other firebases commanding officers were forced to order formal company formations, where the men would be officially ordered to swallow the pills. The commander would then inspect his men to ensure that they had followed his order. At our firebase, the medics would simply give the pills to the men, who were expected to take them on their own.

There were infantrymen in our battalion who refused to ingest the pills. For some it was because of the side effects; for others it was an honorable way to get pulled from combat duty. Any infantryman sick with malaria would be sent to a safe Army hospital in the rear areas for treatment, no matter how long it took. All hospital time was considered "good time," which meant that the time counted toward your in–country year tour. Most of these men probably did not understand that malaria was a lifelong disease that could flare up years or even decades after the original infection. My solution was to break the pill in half and swallow a half one day and the other half the next day. My strategy significantly reduced, but did not eliminate, the cramps or the diarrhea. But it worked. Even though I was bitten many times by mosquitoes in Vietnam, I never became infected with malaria.

My worst encounter with Vietnam's mosquito hordes occurred one night when I was in the Plain of Reeds with my battalion. We had chased the VC all day, but they escaped into Cambodia. After I was ordered to kill trees south of our position (see Chapter 11), I settled down for the night on my air mattress. Before trying to sleep I dutifully applied the mosquito repellent to my face, hands and legs.

My sleep was terribly restless that night in the mud of the Mekong Delta. My nightmare was of rats biting me on the neck like vampires. I awoke in a sweat, felt pain in my neck and discovered, to my horror, that one side of my neck was completely red and swollen out over an inch. My fear was that a rat had actually bitten me; I rushed to find a medic. The infantry medic I found was thorough and sympathetic. After a complete examination, he reassured me that my symptoms were not caused by a rat's bite, but rather the result of being bitten by mosquitoes. Dozens and dozens of bites on top of bites on top of earlier bites had caused the swelling on my neck. Apparently I missed that one side of my neck with the mosquito repellent; subsequently the mosquitoes had attacked that one bare spot in force. I was relieved as the medic applied calamine lotion to my neck. In a few hours the swelling was gone, but that side of my neck itched for days.

14

Reconnaissance Platoon Tragedy

During civics classes in my suburban Pittsburgh high school, we studied the evil of Russian communism and a bit of the history of the murderous Russian dictators like Joseph Stalin, but nothing about the Chinese or Vietnamese versions of communism. During my early college years of 1964 to 1966, I watched President Lyndon Johnson on television tout the "Falling Domino Theory" of why we had to fight in Vietnam. The concept was that if Vietnam fell to the communists, the fall would cause other countries in the region to also fall. His speeches were augmented by vivid graphics. The most common one was a video of a row of dominos, each one labeled with the name of a country in Southeast Asia. The domino of Vietnam would fall and hit the next country domino and knock it down, which would in turn knock down the next country domino and so on, until they all fell. Another video displayed a world map, with Russia in a bright red ink that would bleed down into China and then Vietnam. After Vietnam turned red communist, the red ink would dramatically bleed down into the countries of Cambodia, Thailand, and Laos, finally engulfing all of Southeast Asia, even into Indonesia and Australia. For me, the graphics were all too simple and theatrically overdone, plus I already knew that Johnson was a liar. Fully expecting the war to be over by the time I graduated, I was not worried. I never thought that I would fight in the Vietnam War.

Of course I was wrong. When I arrived in South Vietnam early in 1969, I was neither emotionally invested in the war nor sympathetic to the South Vietnamese government. As far as the average infantryman, he probably knew less about communism than I did. I recall reading about a Navy medic who wrote a report about treating a badly wounded Marine who asked the medic, "What is communism?" and then died in his arms. In one of my letters to my wife, I described the average American soldier's attitude about fighting the war: "They don't like it, they are fighting just to stay alive and they learn quickly."

During those first weeks in country, my primary emotion was fear and shock at the poverty and squalor associated with living in a third world country. While being driven from the 9th Division's headquarters in Dong Tam to my first assignment in Tan An, I saw piers over waterways that ended in outhouses, which emptied the human waste directly into the water. Nearby downstream, Vietnamese women washed their clothes and filled cooking pots while children swam in the same water. There were small Vietnamese houses constructed completely out of flattened Coca Cola cans. Even though I saw mainly the silver aluminum bottoms, there was enough of the distinctive Coke red color to recognize them. Other houses were constructed out of the thick cardboard that encased American C-ration boxes. Yet these were the McMansions of their villages; all the other dwellings were flimsy thatched-roof houses. In the towns, homes were three-walled shacks constructed out of tin sheets and open to the street. Truck drivers, dumping our Tan An base's garbage at dumpsites, would have to drive carefully to avoid the mobs of Vietnamese civilians who were fighting among themselves to be the first to scavenge in our garbage.

New in country, my goals were to control my fear, to perform my duties, and to keep myself as safe as possible in order to survive my yearlong combat tour. I think the vast majority of American soldiers fighting in Vietnam shared these goals. I know they were the goals of 95 percent of the enlisted men I talked with during my tour. But after weeks and then months of being shot at with deadly automatic weapons like the infamous AK-47, being mortared frequently and occasionally being attacked by large caliber rockets, some of my fear evolved into anger. I was mad at these Vietnamese communists for trying their best to kill me. I felt anger, a natural emotion given the circumstances, but not hate.

However, I cannot forget the day my heart hardened and I began to hate the Vietnamese communists. First, background information: the 2nd of the 60th Infantry Battalion consisted of four main-line infantry companies, the maneuver units who did most of the direct fighting. The battalion also included a smaller elite unit of 20 soldiers known as the Reconnaissance Platoon, whose primary mission was to gather intelligence about enemy activities, not to directly engage the enemy. Yet periodically their missions deteriorated into deadly firefights when the communists discovered the platoon. They were respectfully nicknamed "Recon" and

14. Reconnaissance Platoon Tragedy

were issued specially designed camouflage uniforms that only they were entitled to wear. Their camouflage uniforms were heavily striped compared to our standard basic olive-green jungle fatigues; this pattern was called "Tiger Stripes." One knew that any soldier wearing Tiger Stripes fatigues was a courageous man involved in extremely dangerous missions.

Our Recon platoon was led by an infantry lieutenant; there was also a FO artillery lieutenant assigned to the platoon. Before an operation he would always appear in his Tiger Stripes at the TOC to discuss with me his upcoming night mission. We would plot on my wall-mounted AO map his expected night locations. Then he would generate a few potential target positions that the Viet Cong could possibly use in attacking his platoon. He would always ask, not order, me to pre-arrange safety clearances for these potential targets instead of the standard procedure of clearing a target only after the FO radioed it in. The Recon FO was an intelligent and personable man. He and I never really became friends, but he was always friendly and appreciative of my efforts to help ensure his and his men's safety. I liked the man and respected his intelligence and bravery. Only one other FO artillery officer ever came to me to set up this type of pre-clearance target strategy.

Because of my admiration for the Recon FO, I always complied with his requests. In fact, I would go the extra mile for him. I used my contacts with friends in the howitzer batteries to have FDC work out the firing calculations beforehand. Then I asked the "gun-bunnies," the men who manned and fired our howitzers, to pre-aim one howitzer at the most likely Recon target. This pre-aiming ensured the fastest artillery fire on the enemy, if the Recon platoon were actually attacked. Even if that first 105-mm round was not on the exact position of the Viet Cong attack, it would allow the Recon FO to quickly radio in the necessary adjustments, like "right 200 meters, up 100 meters," so that the next rounds would fall directly on the enemy. My gun bunny friends would keep that one howitzer pre-aimed all night, only moving it if ordered to fire the gun on another active target.

To fully understand the horrible night in question, it is essential to understand that the unique mission of our infantry battalion was to work hand-in-hand with what was known as the Brown Water Navy. Not only did the American Navy fight the Vietnamese communists with destroyers on the deep blue waters of the South China Sea bordering Vietnam, it also fought them on the shallow brown rivers, tributaries, and canals found in the Mekong Delta. Our Navy patrolled these brown waters with a variety of heavily armed smaller craft. One such vessel was a shallow-draft troop carrier that looked like the World War II Higgins boat of D-Day fame. It was a rectangular vessel whose one end consisted of a large steel gate that could be lowered to provide a ramp for the troops to quickly debark from to assault an enemy's position.

All of our battalion's infantry line companies rotated duty with the Brown Water Navy every few weeks. The infantrymen considered these assignments more desirable than their usual foot patrolling in the Mekong Delta's mud, because the assigned company would live with the Navy on docked troop ships. Although I never went on a Brown Water Navy mission, the grunts I talked with bragged about the clean, comfortable, air-conditioned living quarters as well as the delicious food the Navy provided. They loved the fact that on any assault patrol, two heavily armed Navy vessels always accompanied their troop carrier boat. They especially loved the one ship that looked like the famous Civil War Monitor. It was a shallow-draft vessel with a clear deck except for one large movable armored turret on the middle, which housed a 5-inch rapid-fire cannon. This gunboat would deliver knockout firepower on the Viet Cong communists while our infantrymen assaulted the enemy position on foot.

Recon's fateful mission that awful night was to search one of the many islands that dotted the middle of the Mekong Delta's wide rivers. Most of these islands were located in remote areas away from population centers. Because of their isolation and the surrounding water barrier, the islands were often used as Viet Cong base camps. These islands were considered dangerous because they were occupied and controlled by the Viet Cong or NVA.

As usual, the Recon FO came to the TOC to pick a few potential targets to pre-clear on the VC island that Recon had been ordered to attack that night. I worked up safety clearances on his target picks and had one howitzer aimed and ready to fire at the most likely target. Later, the undermanned Recon platoon boarded one of the Higgins-type troop carrier riverboats. A Monitor-style gunboat accompanied it. The Brown Water Navy sailed their little armada to the target island.

Tragedy struck early, according to the story an officer told me the next morning. Once they reached the island, the troop carrier made a sharp 90-degree turn so that the vessel could power directly onto the island's banks. There was no enemy gunfire, so standard procedures were followed. The boat plowed onto the island's bank and then dropped its heavy steel ramp directly on the thick vegetation.

Unbeknownst to any American, the Viet Cong had built booby-trap defenses around their island base camp by stringing nearly invisible trip

14. Reconnaissance Platoon Tragedy

Typical thick jungle on the banks of one of the Mekong River's tributaries.

wires connected to rocket propelled grenades (RPGs). Russia and China had supplied the North Vietnamese communists with thousands of these cheap but lethal weapons. Designed as a one-man shoulder-fired weapon, the RPG fires an explosive shape charge from a tube launcher. The grenade is a triangular rocket that, once fired, forms an extremely hot plasma jet of copper that can penetrate thick steel armor. Today our enemies in the Middle East are still using these grenade launchers because they are cheap, abundant and deadly.

The riverboat's falling ramp hit one of these trip-wires, triggering an RPG that fired directly into the vessel. The rocket grenade cut through the Recon platoon standing in the hole, hit the steel at the back of the vessel and exploded. The explosion produced thousands of flying steel fragments that lacerated the men, killing my favorite FO lieutenant and the rest of his 14-man platoon in one horrible blow.

I was struck numb by the tragic news. I could not believe, I did not want to believe that the Recon lieutenant and his entire platoon were dead. The Recon tragedy was our battalion's largest daily loss of American lives during my yearlong tour. Everyone in the TOC was depressed over the bad news. My heart hardened that day. Before then, my primary emotions in Vietnam had been chronic fear and a growing anger, but now they included hatred. I began to hate the Vietnamese communist soldiers. From that day onward I felt an intense drive to kill them. When our battalion mounted offensive missions, I passionately hoped that they would succeed. When our artillery fired on a target, I hoped that it would kill the communists.

An example of my emotional change was demonstrated on Harassment and Interdiction (H&I) artillery fire missions. On slow nights when there was no contact with the enemy, the artillery battery's commanding officer could order H&I fire. The purpose of these H&I fire missions was to keep the Viet Cong and NVA off balance and insecure at night. Sometimes he would give the guns two or three targets based upon recent intelligence reports of enemy movements; other times he would simply pick random targets to fire on. The goal was that the enemy could never know if his night maneuvers or campsites would suddenly suffer destructive artillery fire. As usual, I would have to double-check my maps to ensure that these new target coordinates were not near friendly troops or Vietnamese villages. If they were not, I would give the battery the appropriate safety clearances and record the date, time, target coordinates and clearance initials in our logbook. In general, my sense was that these harassment fire missions were ineffective. The Viet Cong and NVA did not appear to be intimidated. There was no evidence that they were limiting their night maneuvers or attacks.

However, once in a while evidence would demonstrate that our harassment fire had been effective. Occasionally, infantry patrols would stumble upon a campsite littered with torn apart enemy bodies or find enemy bodies that had been obviously struck by artillery fire in rice paddies. The grunts would radio back the coordinates of their find to the artillery batteries and to the infantry TOC. At that point I would be ordered to examine our logbook record to discover if we had fired H&I on the reported coordinates. Most of the time I discovered that one of our howitzer batteries had fired on the reported coordinates during a time period that made sense in terms of the condition of the bodies. The artillery commanding officer (CO) was always delighted when I reported my positive findings, because these enemy bodies added to the batteries' body count total. I never saw a wall chart in the artillery section like the infantry's Body Count Graph, but I know that the artillery CO was required to keep such a tally and report the totals up the chain of command to Brigade Headquarters. There probably was a Body Count Graph at brigade headquarters comparing the various member batteries.

I never celebrated these body count reports of H&I success outwardly, either before or after the loss of our entire Recon platoon. But I must admit that after that horrible event, I would inwardly feel a positive sense of satisfaction that I had contributed to killing Vietnamese communists. It was not that I had somehow shortened the war or hastened my return home from the war zone. No, it was purely the euphoria that American combat soldiers in Vietnam called "Payback." Grunts expressed the concept in their mantra: "Revenge is sweet, but Payback is a Mother Fucker."

15

Burning Shit

By far the most disgusting job I was ordered to do in Vietnam was disposing of our human excrement, or as everyone, including all the officers, called it, "burning shit." In American firebases the toilets were wooden outhouse structures of various sizes, but they all had the same basic design. The two-hole outhouse behind our TOC was a typical one. A soldier would have to climb up two steps to open the door, which did not have a half-moon design. Once in, he found two raised toilet seats attached to the top of a wooden boxed frame. He would have to step up to seat himself on one of the standard toilet seats. Underneath each seat, inside the boxed frame, was one half of a 50-gallon drum that had been cut to fit into these boxed frames. There were rolls of toilet paper sitting on the box, as well as a big can of lye with a scoop in it. The soldier would sit on one of the seats, do his business, wipe himself clean with the toilet paper and then drop two or three scoops of lye on top of his deposit.

Only low-ranking enlisted man were assigned to the repulsive shit-burning detail. All of the battalion's headquarters' enlisted men below staff sergeant were assigned to service the outhouse behind the TOC. The sergeant major posted a monthly schedule that listed the name of the soldier assigned to the duty for every day of the month. His schedule worked out that I had to burn shit about once every two weeks. When my dreaded day came, I retrieved the required tools from the TOC. They were extra large asbestos gloves, a thick broom-handle–type rod with one of its ends badly singed, and a five-gallon metal container of diesel fuel.

The actual job consisted of a number of steps. First I would put on the gloves, walk to the back of the outhouse, lift the hinged wooden flap that covered the back of each toilet and pull out the 25-gallon drum that was filled with the excrement. Grasping the handle hole that had been cut into the drum, I would drag the drum about 150 yards along our firebase's

15. Burning Shit

metal walkways to a remote section of the base. This one area had been designated as the dumping ground for the burnt crap. I repeated this dragging with the other excrement-filled drum from the outhouse's other toilet. Then I had to pour approximately a gallon of diesel into each of the drums and set them on fire. The resulting blaze produced clouds of pitch-black smoke and a foul effluvium.

The worst part of the job was stirring the crap and diesel with the broom handle rod to ensure that all of the waste was burnt. To do so required that every few minutes I endured the heat and the odor of the nauseating fire to stir the mess. It usually took about 30 to 40 minutes for the day's accumulation of feces to fully burn. Once the waste had burned to ashes, I was then required to dump the waste over the end of the last walkway. That rice paddy was nearly filled with the accumulated years of a compound of mixed burnt and not so burnt crap ash. It too had a peculiarly vile stench. Of course, after the crap was dumped, I had to drag both half barrels back to the outhouse and reinsert them into the box frame under their respective toilet seats.

At its best, burning shit was a sickening job. Yet there were complications that made the work even worse. Too often the other soldiers who shared the burn work did not perform as complete and efficient a job as I strived to do. The most common inadequacy was that they would not completely incinerate all the excrement; consequently, the returned half barrels were not empty when replaced under the toilet seats. A couple of times the result was that I had to pull out and burn half barrels that were overflowing with crap, which was harder to light and took longer to burn. The worst problem was when a man would simply not perform his duty. Again, too often I would have to deal with two or three days' accumulation of human excrement. One ugly sight branded into my memory is when I pulled out the half barrels and found them filled to the brim with slithering white maggots because the accumulated waste had not been burnt for days.

Once in the wet season I was dragging a fully loaded half-barrel of crap when I slipped on the wet metal walkway and fell into the piles of accumulated half burnt shit underneath. At least I had my boots on, not my usual flip flops, yet I still wound up sinking into a revolting mixture of crap and ash. It nauseated me so badly that I almost vomited. That night I spent over an hour trying to clean off my boots, even cleaning out the small lacing holes that had filled with crap. I was never able to completely eradicate the smell; those boots released a shitty odor for weeks.

Soon after the maggot incident, I complained to the battalion's sergeant major that some of the assigned enlisted men shit burners were not doing their duty. I mentioned that the supervision of this work detail was too casual, allowing these slackers to avoid their responsibilities. Finally, I suggested that the best solution to the problems we were experiencing was to hire a Vietnamese civilian to do the burning. I knew this solution was working at a number of the Delta's other firebases. A South Vietnamese civilian performing the backbreaking work of planting and harvesting rice earned about 35 cents a day in the field. The other bases were paying their civilian burners one dollar a day. The Vietnamese clamored for these jobs. We would have no trouble finding a civilian willing to do the job.

At that time, I did not know that the sergeant major was married to a Vietnamese woman; thus his attitude toward Vietnamese civilians was much different from mine. We argued back and forth; I do not recall the exact details of our argument, but I remember thinking that he really did not understand how disgusting the job really was. I do recall that he argued that I was not much of a man. Then he claimed that he was going to prove that he was a better man than I by burning the shit himself. His reasoning surprised me; I kept quiet to allow him his victory. And yes, he did pull out the half barrels and drag them to the dumpsite to burn, while I went back to my bunker to celebrate by smoking a joint.

An hour later, a few of my bunker mates returned to our bunker and were surprised to see me relaxing on our deck. They said that they knew it was my day to burn the shit and that they thought they saw me at the dumpsite burning the half barrels. Convinced that it would be funny to harass me, they shouted and hooted taunts at the figure stirring the burning crap. They had shouted insults like, "You go, boy, burn that shit! You stir it good, boy!" They did not realize that the sergeant major was the target of their mean-spirited jeers.

Within the next few days, someone, possibly the harassed sergeant major, hired a Vietnamese civilian to burn the excrement. I never had to be on that putrid burning detail again. I do not know how much they paid the man, and I do not remember having to donate any money to his payroll. He could have been paid from the battalion's slush fund. I often wondered if my argument with the sergeant major had anything to do with the pleasant turn of events, or if the results were due to my buddies' yelling at him. All I know for sure is that my Vietnam tour was much less disgusting after the new hire. And think about it: we were contributing badly needed hard cash into the South Vietnamese economy.

A previous development in Dong Tam demonstrates the terror that plagued many infantrymen in the Vietnam War. During my first week orientation to Vietnam at the 9th Division's base camp, I saw a soldier

15. Burning Shit

whistling as he pulled two half barrels of shit. He had a wide smile even though he was doing such a repulsive job. Surprised by his happy demeanor, I asked a sergeant about the guy. He told me that the happy shit burner was a draftee who had been assigned to the infantry. He was so completely unnerved by combat patrols that he started petitioning his commander for a non-infantry assignment. Many grunts had the same wish, but this soldier was unusually loud and persistent. Eventually, his commander tired of his whining and decided to test him. He offered the infantryman a permanent shit-burning assignment. The grunt was so terrified of infantry fighting that he accepted the offer, even though it meant that he would be burning the entire base's shit every day, all day, for the rest of his yearlong Vietnam tour. It was weeks later when I spotted him smiling and whistling, while doing the foulest job I knew of in the Nam.

16

Fragging

During the Vietnam War the term applied to the deliberate assassination of an officer or a non-commissioned officer (NCO) was "fragging." The name came from the American M67 Fragmentation Grenade, which was a 14-ounce spherical steel-bodied explosive, designed to be baseball sized so that a soldier could comfortably throw it. When the grenade exploded, hundreds of sharp metal fragments would shoot out in all directions. It was often used in the Vietnam War for assassinations because its explosion destroyed all evidence that could pinpoint the murderer. In the Vietnam War the term fragging generalized to include any attempt to assassinate a higher-ranking individual, no matter what the weapon.

My contention is that fragging attacks on non-commissioned and commissioned officers were more frequent in the Vietnam War than the U.S. military officially admitted. In Vietnam War battles, so many soldiers died from grenade fragments that frequently a fragging attack was not even recognized as a deliberate assassination. My argument is that the military misdiagnosed such attacks as routine combat deaths. I also suspect that there may have been an effort to keep the real number of fraggings hidden, so that the totals would not negatively affect officers' morale and in order to limit copycat murders. There were four attempted fraggings in our infantry battalion during my ten months at the Tan Tru firebase.

The first incident occurred early in my tour. The 1960s Black Power movement that arose in the States spread into Vietnam. Four Black infantrymen from one of our front line companies thought that there was racism in the company. They claimed that Black infantrymen were being assigned to the most dangerous missions. To demonstrate their anger, they marched into the company's headquarters and threatened their company commander with a loaded M-60 machine gun, loaded M-16 assault rifles and hand grenades. Obviously, they lacked adequate rear security, because

16. Fragging

while they were asserting their complaints and demands to the captain, a lieutenant from the company snuck up behind them and shoved a 45-caliber pistol into the head of their leader, who was the one brandishing the M-60 machine gun. The lieutenant forced the infantrymen to give up their weapons, thus ending the mutiny.

I am not sure why the culprits were not arrested and punished; instead, all their gear was quickly packed up for them. Then they and their gear were thrown on four helicopters that transported them to other units located at the four farthest corners of South Vietnam. The colonel must have decided that the best response was to separate them rather then publicize the incident with a court-martial.

One nasty character who stimulated a number of fragging attempts was the artillery battery's sergeant major. He was the type of NCO who reveled in wielding the power he had over lower-ranking soldiers because of the Army's authoritarian chain-of-command system. To me, he was one of those military losers who enjoyed asserting power over brighter, more talented men, knowing full well that, given his limited talents, he could never do so in the civilian world. He continued to harass everyone he could throughout his tour. He punished men for minor infractions like not having nametags sewn on their fatigues. He forced all enlisted artillerymen to have their hair cut short every two weeks. On my last day in the unit he ordered me to have my hair cut as short as possible so that I would return home with a shaved head. In the United States during the 1960s and early 1970s, long hair on a man was an important cultural issue and status symbol. There was no Army regulation requiring such a discharge haircut; it was only his meanness.

This sergeant major's harassment of enlisted soldiers in the battery was so harsh and unnecessary that the artillerymen placed a bounty on his head. Whoever killed him, American or Vietnamese, would receive hundreds of dollars. The bounty grew and grew throughout my tour into over a thousand dollars. Whenever the sergeant major persecuted lower-ranking soldiers, his victims would throw more money into the assassination bounty. Even if no one ever collected the money, it gave the harassed low-ranking soldiers a feeling of owning some power while caught in such an official subservient position.

But for some that bounty outlet was not enough. One night someone threw a grenade into the sergeant major's sleeping bunker's air vent. It bounced off of the recessed screen covering the vent, exploding just outside the bunker. Thinking that the explosion was caused by an enemy mortar attack, the sergeant ordered the mortar platoon to fire at counter-mortar targets. I was on duty at the TOC that night and heard the unsanctioned outgoing mortar fire. At once I phoned the mortar pit to discover what was going on. My concern was why were they firing without first obtaining the required safety clearances. I was told that the artillery area was under mortar attack.

In response I alerted the counter-mortar assessment team, whose job was to assess damage and to estimate the direction and range of the enemy's fire from a survey of the blast hole and any mortar fragments left. They radioed back to the TOC that the explosion was caused not by a mortar shell, but by a grenade. I quickly radioed a cease-fire to our mortars. The best part of this incident was that afterwards I was able to confront, not harshly I must admit, this hated sergeant major over his breaking of our artillery protocol by ordering mortar fire without the required safety clearances. Surely this justified confrontation did not endear me to him.

The sergeant major did not change his behavior after the grenade attack; instead, he ordered an escape hatch built into his sleeping quarters. Previously he had lined the room with cheap wooden paneling. In his new remodeling, he removed one siding panel and had an opening cut out behind it. Then he had a one-man mini bunker built on the other side of the opening. Finally he had the panel rehung without being nailed to any supporting studs. His plan was that, if attacked again, he could escape harm by jumping against the unsecured wooden panel, which would then collapse into his mini-bunker. As far as I know, during the rest of my tour he was never forced to test out his escape plan.

One night a few weeks later this psychopathic sergeant major was sitting alone, watching a pornographic movie projected on the outside wall of an artillery bunker. During the movie an assailant stepped out of the shadows and fired his M-16 assault rife at the sergeant major. He fired a ten-round clip in full automatic mode, but his aim was inaccurate. The first round struck the bunker wall a few inches above the sergeant's head. Then the assailant's M-16 rifle did what they always did when fired in the full automatic mode: the speed and the power of the bullets made the rifle's barrel rise with each round. Inspections of the bullet holes demonstrated that the rifle had drilled a pattern up and to the right of that first round.

My bunker mates and I knew that this climbing pattern of bullet holes proved that the failed assailant was an artilleryman, because an infantryman would never allow his M-16 barrel to climb like that when firing in full automatic. In the Vietnam War, infantrymen learned that, to prevent their barrel from climbing, instead of the conventional below-barrel grip they had to grip the barrel from above. With this unconventional grip the

16. Fragging

infantryman could push down on the M-16 barrel when firing full automatic, a maneuver that kept the barrel stable and the bullets on target. No soldier was ever arrested for either of these fragging attempts on the artillery battery's sergeant major. In fact, I do not recall any actual investigation into these assassination attempts; it seemed that they were accepted in our battalion as a normal cost of war.

There was, however, one successful indirect fragging in our battalion during my Vietnam War tour. To fully understand the incident requires considerable background information. It starts with the body-count mentality of the American military during the Vietnam War. Since it was not a conventional war, the usual measures of military success like cities liberated or regions conquered were not applicable. Under the influence of Secretary of Defense Robert McNamara's systems analysis approach to running the war, U.S. commanders morbidly settled upon the number of enemy dead as the primary measure of success in the Vietnam War.

This system analysis metric may have worked for Mr. McNamara when he was running the international Ford Motor Company, but it was not an effective strategy for running an unconventional war like Vietnam. This reality is especially true when one considers the attitude of the communist commander Ho Chi Minh. It is reported that when his own North Vietnamese Army commander, General Giáp, complained to Ho about the tens of thousands of NVA soldiers killed during their attacks on the Khe Sanh Marine base in 1968, Ho is said to have replied sarcastically, "What do you want me to do, awaken the dead?" The number of his peasant soldiers killed meant little to Ho Chi Minh. Historians find it difficult to establish the number of North Vietnamese soldiers killed during the Vietnam War because the communist leadership did not keep count. Those human deaths were inconsequential to them. In a bizarre twist, the American military were the only ones counting the enemy deaths.

Our military's body count mentality led to such grotesque concepts as the kill ratio metric, which was the ratio of American dead to enemy dead after any Vietnam War battle. This ratio was considered the ultimate measure of an Army unit's combat effectiveness. The body count mentality also led to the morbid chart that hung on our battalion commander's office wall in the TOC. It displayed, in bar graph form, the total number of enemy kills for each of the battalion's four infantry line companies. The line company with the lowest number of enemy kills in a month would be berated and sometimes punished.

During the first few months of my tour, our battalion's Charlie Company was commanded by a captain who did not agree with the aggressive nature of the American military in Vietnam. When Charlie Company was ordered to stage an airmobile assault, after they had landed their captain would order his company to march to the nearest overhead cover in the jungle. Then he would have his men form a circular perimeter around him and his headquarters unit. He told his infantrymen that, as far as he was concerned, their primary mission was to protect him. In return, he would take care of them, ensuring that they were fed the best hot food, that they would receive their Rest and Recreation leave (R&R, a week long vacation from the war), and that they were paid on time.

Charlie Company's commander would then radio in false maneuver reports to the battalion commander, who was flying around overhead in his C&C helicopter. These false reports would state that Charlie Company had searched 2,000 meters in one direction and then had searched 2,000 meters in another direction and so on, until at the end of the day the company had maneuvered back near their drop zone. These reports were all lies, but then that was what America's Vietnam War was based upon: lies. The infantrymen in Charlie Company loved their captain and his safety-first strategy. Eventually, the captain's combat tour ended.

By that time our battalion's commander was upset by the fact that Charlie Company had consistently recorded the lowest number of enemy kills as displayed on his office's body count wall graph. He decided to remedy this problem by assigning to Charlie Company a new captain who was aggressive and eager to find and destroy the enemy. He soon found such a hard-charging captain. The colonel appointed the new company commander with the explicit expectation that he was to increase Charlie Company's enemy body-count numbers.

Afterwards the new commander gathered the men of Charlie Company to inform them of his aggressive fighting philosophy. From that day on they were going to hunt, engage and kill the VC and the NVA so savagely that the enemy would come to know and fear the company. He cited the need for the company to develop a strong esprit de corps. As a symbol of their new belligerent spirit, he ordered all of the Charlie Company infantrymen to wear red bandanas around their necks over their camouflage uniforms, even when they were out in the bush on search and destroy missions.

The infantrymen of Charlie Company were upset by the order; they all understood that wearing a red scarf in the green jungle was suicidal. The red bandanas would pinpoint their location to the enemy. To do so would be like wearing a large red bulls-eye target on their backs. Yet what could they do? Disobeying such a direct order in a war zone was a transgression that would be severely punished. So they wore their hated red bandanas and were extraordinarily fearful.

Roughly two weeks later, Charlie Company was ordered at the end of one of their search and destroy patrols to walk back to our firebase. They were following this order when they came upon a wooden footbridge that spanned a canal. That canal and its footbridge were well known to the men because it was close to our firebase. Every one of Charlie Company's infantrymen knew that the footbridge was extensively booby-trapped with explosives. The infantrymen of Charlie Company felt desperate. Desperate men do desperate things; they felt that they had nothing left to lose. So none of the men warned the captain of the booby trap. As he trotted over the mined bridge he triggered a booby trap, igniting an explosion.

Early that night I went to our enlisted men's club for a drink. I had not yet heard about the day's tragedy. Inside the club I spotted Charlie Company's medic drinking heavily at the bar. His elbows were on the bar and his head was down. He looked depressed and upset. I sat down next to him and asked him what was wrong.

Sorrowfully the medic recounted the details surrounding Charlie Company's tragic patrol that day. The exploding booby trap had severely wounded the new company commander, puncturing both his chest and lungs. He suffered what the Army called a sucking chest wound. The captain's chest had been torn open, exposing his lungs. As he struggled to breathe air, he sucked blood into his lungs instead. The new commander was drowning in his own blood.

This experienced medic had treated soldiers with sucking chest wounds before. Those challenging experiences had taught him to carry, along with his standard Army-issued medical gear, his own rubber tubing. After diagnosing the captain's plight, the medic forced the rubber tube down the captain's throat into his lungs. He then proceeded to suck out the blood from the captain's lungs with his own lungs and mouth. He would spit out a mouthful of blood and then suck out another mouthful. He repeated the procedure over and over again. Sadly, the captain's blood poured into his lungs faster than the medic could suck it out. Before the medical evacuation helicopter arrived, the wounded captain drowned in his own blood.

I admired the courage of the medic's desperate attempt to save his captain. When his eyes filled with tears, I tried to console him. He then told me that after the death, every Charlie Company infantryman on that patrol walked up to the body, pulled off his red scarf and threw it down on the corpse. The hard-charging captain lay there dead with nearly 100 red scarves piled all over his body.

The most outrageous attempt at fragging occurred in December, when our battalion was assigned a new commanding officer. Our previous colonel was a true leader. He led from the front; his men could see him. During battles he landed his C&C helicopter to consult with company commanders. He shared the hardships his men had to endure. When the battalion was forced to stay overnight in the field, the colonel was out there with his men, sleeping in the mud and eating C-rations like we all did. All the men in the battalion respected the colonel for his genuine leadership skills.

Our new commanding officer did not agree with our previous leader's share-the-pain philosophy. The new colonel had previously been assigned to the Pentagon; he demanded the same pampered lifestyle in Vietnam. He always wore starched jungle fatigues accompanied by spit-shined boots. He looked out of place. He ordered me out of my t-shirt, shorts and flip-flops; he insisted that I wear my complete jungle fatigue uniform, including boots, when working at the TOC. No matter how fierce the battalion's firefights were, the new colonel always flew back to our firebase for lunch. Every night, no matter if his men were forced to sleep in the field that night, he flew back to the firebase to sleep on his clean sheets in his air-conditioned quarters.

This colonel possessed an elite standoffish command style. He flew over the battles in his C&C Huey, ordering his companies to maneuver here and there as if they were chess pieces on an enormous battlefield chessboard. In my December 14 letter to my wife, I wrote: "It makes me frustrated to see how he is putting the screws to everyone of the grunts." He never landed his helicopter to talk to his field commanders or to observe the battlefield from their perspective. He did not understand that the battlefield appears much different from the air than it does to the infantrymen on the ground embroiled in the fighting.

Moving a company from one position to another appears easy from the perspective of a helicopter flying over the jungle or rice paddies, but on the ground the reality is much different and more difficult. Most of the time, the maneuvering infantry companies had to cross through energy-draining mud, or had to cross over swift waterways without the aid of rafts, or had to cut their way through thick primeval jungle. Yet the colonel was always criticizing his company commanders for not moving fast enough. Most of the infantrymen learned to hate this new gung-ho colonel.

This hatred grew to the breaking point. Soldiers can become so desperate that they think that they have nothing left to lose. As Kris Kristofferson wrote in his song *Me and Bobby McGee*, "Freedom's just another word for nothing left to lose." Desperate men with nothing left to lose possess the freedom of no moral restraints. They can become so angry that they seek to act out that anger against whomever they think is the cause of their desperation. This no-restraints attitude was expressed by the commonly verbalized question asked by soldiers contemplating desperate behavior: "What are they going to do, shave my head and send me to Vietnam?"

Someone, probably an infantryman, set up a tripwire low across the threshold to the colonel's door. The tripwire was connected to a hand grenade, set to explode if the tripwire trigger was touched. The colonel did trip over the wire, but the hand grenade was a dud; it did not explode. Officers speculated that the grenade was deliberately made a dud in order to only frighten the colonel. This argument was made null and void the next night, when the colonel noticed a new trip wire set up across his door, connected to another live hand grenade.

The next day the colonel ordered that an armed soldier guard his door 24 hours a day. He also ordered that a roll of concertina barbed wire be strung out around his sleeping quarters. Now I must clarify that both the armed guard and the barbed wire were meant to protect the colonel, not from our communist enemies, but rather from being killed by his own soldiers. This is another example of the upside down, surreal never-never land of America's Vietnam War. It was as if we were living in our own nightmare, except that it was far too real.

17

The Food

Any Vietnam War memoir would be incomplete without including a chapter devoted to the food our military fed its troops. The majority of my meals during my yearlong Vietnam War tour were taken in the 2nd of the 60th Infantry Battalion's mess hall, affectingly called the grunt mess hall. Three meals a day were prepared and served cafeteria style by Army cooks supervised by a senior NCO head cook. Our grunt mess was a large wood-frame structure built on a deck resting on wooden pilings. The mess hall housed the kitchen, a serving line and about 25 round tables that could seat six soldiers each. It was not the cleanest cafeteria I ever ate in; consequently, it attracted rats, even during the day. Its informality attracted me. Most days I dined there in my usual Vietnam uniform of the day: cut-off jungle fatigue shorts, a camouflage T-shirt and rubber flip-flops.

Most of the meals served in the grunt mess hall were greasy and forgettable. I do remember my many breakfasts, which always consisted of scrambled powdered eggs and toast. Since I had just completed an all night duty shift at the TOC, I did not drink the coffee because I planned to sleep afterwards. But I do recall a scene of a cook stirring his fresh coffee in an immense 25-gallon aluminum pot. Milk was never available at our mess hall. It was easier for me to acquire a glass of beer there than a glass of potable water. Instead I was forced to drink the mess hall's cherry flavored Kool-Aid, which had a metallic iodine taste from the purification tablets added to the water. The Kool-Aid was stored and served from an Army issued olive-green 50-gallon bladder hanging in the mess hall. To this day I cannot stand the taste of Kool-Aid.

Mostly I would sleep through lunch and then awake for dinner. The only dinner meals I do recall were the few times the cooks grilled steaks outside the mess hall. All the other dinner meals deserved to be forgotten. After a few months in country I developed chronic stomach discomfort

17. The Food

that I attribute to a combination of the greasy food and the overall stress of the war. It was never severe enough to seek medical treatment, but the stomach pain persisted for months. Basically, I ate the food that was served; my sniper roommate was a much more picky eater. One time he filled his tray up with the usual fare and then threw it at a cook in disgust. He was lucky that he was not prosecuted for his acting-out tantrum.

One incident I observed in the grunt mess hall dramatized America's rich abundance of food compared to the hunger many South Vietnamese civilians endured. One evening I went to the mess hall near the end of dinnertime. I filled my standard brown cafeteria tray with plates of food and sat down to eat alone. As the cooks were closing down the serving line, rather than throwing food into the garbage, they allowed a teenaged Vietnamese boy to partake in the leftover mashed potatoes. The adolescent was the son of one of the dozen or so Vietnamese women that worked on our firebase. I watched in fascination as this thin teenager, who looked deprived, took a cafeteria tray and, rather then using plates, filled the entire tray with mashed potatoes piled high. Then he carried his overloaded tray to a table, sat down, took the salt shaker standing on his table, screwed off the top, and dumped all the shaker's salt onto his pile of potatoes. He did the same thing with the pepper shaker and the sugar bowl. He then mixed the condiments into his pile of mashed potatoes and proceeded to gobble down the entire tray full of potatoes.

One of the reasons I can remember the grunt mess hall's steak dinners is that the cooks would serve us free beers with the steak. The late 1960s was an era of major NCO financial corruption scandals like the one involving kickbacks to the Command Sergeant Major—the highest ranking NCO in the entire Army—from the slot machines installed worldwide in enlisted men's clubs. The free beer with dinner in this era of widespread NCO corruption made me more than a little suspicious. A standard ploy of successful psychopaths is to give victims a little something as a kind of distraction, while cheating them out of much more. Free beer and steaks grilled outdoors by senior NCO cooks; the only time I saw them actually working. Definitely something suspect was happening.

My suspicion is that the senior cooks received Army funds to buy top grades of beef to serve our soldiers. Instead they bought cheaper meat, probably Vietnamese water buffalo steaks, and pocketed the difference. The steaks were a little tough, but when they were accompanied by free beer, who was going to notice or complain? I ate the steaks and drank the beer. It was a tasty meal, a picnic that was much better than the usual mess hall meals. These steak dinners occurred about once a month. I must have eaten nearly one-half of a water buffalo over the course of my tour, while the cooks escaped investigation.

These senior NCO cooks also devised another more obvious corruption scam during the last three months of 1969. One day out of the blue they opened up a pizza parlor on our firebase. Pizza parlor on base! Are you kidding me? We of the lower ranks were thrilled. Their pizza parlor was a small two-room building manned by South Koreans, who now suddenly lived on the firebase. However, the parlor was always supervised by one of the obese senior NCO cooks, who collected the money. The pizzas cost one dollar each; their crusts were paper-thin, topped only with tomato sauce and a little cheese. The Koreans baked them in a standard Army oven.

The firebase's new pizza parlor was a big hit. The line of enlisted men outside the store was always a long one. A few days after it opened, I tolerated the long line to try out the new treat. While standing in line, I observed that just one of the thin pizzas was not much of a meal. So the always adaptable American soldier coped by buying three or four pizzas at the same time and then stacking them on top of each other to make a thick pizza. When my turn came, I ordered two to make myself a pizza sandwich. The oven and the ingredients were hidden from view in the back room. A Korean served me my pizzas, but I had to pay my money to the overweight E-7 cook sitting at the end of the counter. He took my money and piled it into a cigar box full of money that he guarded on a nearby stool.

I have no real proof, but I would be willing to bet my life savings that the senior cooks were stealing the tomato sauce, cheese and flour from the Army's food storage Conex boxes on base. They accepted payments for the pizzas, paid the Koreans a pittance and pocketed the rest. It was quite the successful illegal scheme. I think that these senior NCO cooks made themselves bundles of money. But the pizzas were hot and delicious. No one complained, and apparently the commanders looked the other way, or perhaps they were paid off too.

Since I worked as an artillery liaison to the infantry, I was considered a member of both groups. This provided me the advantage that I was eligible to take my meals in both the infantry and artillery mess halls. The artillery mess hall, located in the artillery section of the camp near the howitzers, was a longer walk from my bunker. It was smaller but cleaner than the infantry mess hall. The food, standard American cuisine, was the same; yet it always seemed to taste better there than the food at the grunt mess hall. Its one big drawback was that the artillery's first sergeant cook insisted that the soldiers eating in "his mess hall" wear full regulation jungle fatigue uniforms. I could not partake in my usual T-shirt and shorts, so I ate there infrequently.

However, there were times when I would take the trouble of wearing my fatigue uniform to eat there, usually when I learned that the menu at the artillery mess included any of my favorite foods. The one day that I was especially grateful for being eligible to eat at the artillery mess was the Thanksgiving holiday. The Army went all out to ensure that every one of its soldiers was served a hot meal of turkey and all the trimmings on Thanksgiving. For the troops in the field, a cook would deliver the hot turkey and all the trimmings directly to the men.

At Thanksgiving the artillery mess was transformed into a homier atmosphere. I remember our Thanksgiving meal fondly. The dining room was decorated with paper turkeys and pumpkins. The tables were all joined together to form one large dining table that was covered with white tablecloths (they may have been white sheets, but no matter). Instead of cafeteria style, Thanksgiving was served home style. All of our dinner plates, glasses and utensils were already laid out around the table. There was also a lovely Thanksgiving-themed centerpiece decorating the table. The two batteries of artillerymen sat around the table in our clean uniforms, with our captain joining us at the head of the table. He started the celebratory meal by saying a short prayer of thanks. Afterwards the cooks, wearing white aprons, served us the meal. We all ate our fill of hot turkey, sweet potatoes and cranberries. There was even pumpkin pie for dessert.

The artillery Thanksgiving holiday event worked for me. I thoroughly enjoyed and appreciated the ritual, the traditional dinner and the camaraderie. I felt totally comfortable eating with these men, both black and white. It was the first time in the war that it felt like we were all in this fight together, not really one big happy family, but certainly a formidable team. Judging by the smiles and laughter, I think the others felt the same way.

When I travelled to Saigon, for whatever reason, I always tried to take advantage of the opportunity to visit the Tan Son Nhut airbase, located in the northeast section of the capital, to dine at the Air Force's mess hall. In Vietnam no American military personnel would be refused service at any of our military's mess halls. The Air Force mess hall was a large, clean, well-lit facility that would look at home on any American college campus. Best of all, it was air-conditioned. It provided a wider selection of delicious food than any Army mess hall. The main culinary attraction for me was that the airbase's mess hall served real milk. I luxuriated in drinking down quarts of delicious and nutritious cold milk in air-conditioned splendor.

Once I ate at a Navy mess hall on the outskirts of Saigon. The Navy mess was better than the Army ones, but the Air Force's mess hall was far better than the Navy's.

The other major source of my nourishment in Vietnam was C-rations or "C-rats," officially named Meals, Individual Combat. C-rations were commercially produced canned food meals designed so that soldiers could easily eat them in the field. They came packaged in a thick cardboard case that contained 12 separate meals, each of these packaged in its own thin cardboard box. Each individual meal box contained a can of meat, another can of fruit or bread and a flat spread can of cheese, jam or peanut butter. It also contained an accessory pack filled with items like coffee, powdered cream, cigarettes, gum and toilet paper. Finally, each case included four small can openers.

The C-ration can openers are named P-38s. In my opinion they are one of the most well designed tools ever manufactured for the U.S. Army. The P-38 consists of a piece of thin steel one and one half inches long, hinged to a half-inch steel pointed hook blade. To open a C-ration can, the soldier pierces the can's lid with the blade and then turns the handle as he moves the P-38's blade around the lid. They are inexpensive, lightweight, portable, efficient, and durable. In Vietnam every American soldier going on a combat operation carried a P-38 as part of his indispensable field gear.

The C-ration meals could be eaten cold in the field, but they were more palatable when heated. Army issued Sterno tablets could heat the meals; however, the most popular method was to use C-4 plastic explosives. On field operations, combat engineers and most infantry companies carried demolition blocks of the plastic explosive. C-4 could also be obtained from Claymore anti-personnel mines. Since C-4 has the texture of modeling clay, a soldier could easily break off a small piece of the explosive and roll it into a small marble-sized ball. He would then place the ball on the ground and light it. The explosive would burn as a white flame with an intense heat, while the soldier held his can of meat over the fire by holding on to the can's bent back lid.

This method of heating C-rations would cook the food faster and more thoroughly than the Sterno tablets could. However, when burning C-4, there was always an inherent danger of toxic fumes and an explosion. The solution was to simply allow the ball of C-4 to burn itself out. Unfortunately, we did have an infantryman in our battalion who did not consider the danger. He stomped on his burning ball of C-4 to extinguish his fire; it exploded, blasting his foot completely off.

17. The Food

The 12 meals packed in a case of C-rations varied significantly in desirability with the troops. Every soldier in the field had his own personal favorites. My favorites were the spaghetti in tomato sauce and the boned chicken meals. My favorite desserts were the canned peaches and the fruit cocktail. Everyone loved the pound cake. For me, the peanut butter spread over the round crackers made an acceptable lunch, especially when paired with a can of fruit. There was also universal disgust for a few of the meals, especially the ham and lima beans (often referred to as ham and turds) and the beans and weenies (official name: beans and frankfurters). Soldiers would negotiate, trade and sometimes fight over their various C-ration meal favorites. Since I did not smoke cigarettes, I used the small packs included in the C-rations to barter successfully with smokers for a desirable item like a can of peaches. Frequently the undesirable meals would be forced upon new replacements. To eliminate the conflicts, one company commander turned all the individual boxes over and randomly mixed them up before he handed them out.

Some troopers evolved into C-ration gourmet chefs when eating out in the bush. Every one of these C-rat chefs stole bottles of tabasco sauce from the mess hall in order to add the hot sauce to their C-ration creations. The mess hall cooks were always complaining of their difficulties replacing all the stolen bottles of tabasco sauce. My favorite C-ration chef was a FO lieutenant of Philippine descent. I was out in the field on an encirclement operation when his infantry company was picked to provide security for our headquarters company. This American officer was dark skinned and small of stature, and he had Asian facial features. He looked like a Vietnamese man in an American Army uniform. The older Vietnamese women, the mama sans, in the nearby village were attracted to him. At dusk on our first day out, he was able to convince one of the older women to allow him to use her hut's cooking stove. The stove itself was made of hardened clay with metal strips across an opening over the fire below. There was another storage cubbyhole for the slender lengths of bamboo used to feed the flames.

This artillery FO chef asked seven of us artillerymen to donate our cans of meat to him. We all opened our cans and gave them to him. He had negotiated with the older woman for about two pounds of rice in exchange for some extra C-rations cans of meat. He then poured all the contents of the meat cans and rice into his upturned steel helmet, which he was using as a cooking pot. He mixed the food and poured a bottle of tabasco sauce into the mix while cooking it all on the open-fire stove. From the hut's doorway, I watched in fascination as he also mixed in salt, pepper, and a bit of Vietnamese spices, while continuing to add more bamboo sticks to the fire. The lieutenant simmered the mix for about 15 minutes. Declaring it ready, he served the hot rice mix to himself and the donors; we all savored the spicy dinner. His exotic C-rations potpourri was the best meal I ever ate while out in the field.

At our firebase, cases of C-rations were not difficult to acquire. There were abundant stores on the base and little accounting of their numbers. Many nights, when a group of my bunker mates and I developed a heightened appetite after smoking marijuana, a couple of guys would go out and procure a case or two of the C-rations. We all would excitedly break open the cases and take out our favorite meals. Most of the time we simply discarded the unappealing ones, such as the ham and turds, into the trash. These rejected discards were one of the reasons why Vietnamese civilians fought each other for the chance to rummage through our garbage dumps: they were hunting for the olive green C-ration cans.

18

Weapons

M-16 vs. AK-47

One of the myths propagated by the anti–Vietnam War left was that we were fighting against ill-equipped Viet Cong guerrillas. The left's propaganda made it seem like the VC were fighting only with sharpened bamboo sticks or bows and arrows. The propaganda was a total lie; the reality was much more savage. Both the VC and the NVA were equipped with sophisticated Chinese and Russian state-of-the-art weapons. Our communist enemies fired grenades, rockets, mortars and rifles in their attempts to kill, not bamboo spikes or arrows.

The best example of their advanced weaponry is the fact that the Vietnamese communists were equipped with the infamous and ubiquitous AK-47 assault rifle as their main battle rifle. AK-47 stands for "Avotomat Kalashnikov" or assault rifle, designed by Soviet designer Sergeant Mikhail Kalashnikov and adopted by the Soviet Army in 1947. It is a gas-operated assault rifle capable of firing a 7.62-mm round in either the full automatic or semi-automatic mode. It is cheap to manufacture, highly reliable, easy to operate, and able to tolerate massive amounts of abuse. It was very effective in Vietnam because its heavy 7.62-mm round was able to penetrate the jungle foliage, whereas the 5.56-mm round fired by our M-16s could not penetrate Vietnam's jungle foliage. Even thin branches would deflect M-16 rounds. Millions of AK-47 assault rifles have been manufactured by a number of different communist countries. Military experts estimate that the AK-47 is the most lethal weapon ever made, when measured by the number of humans killed by a weapon. The rifle has caused more deaths than atomic bombs or artillery. Even today, 50 years later, enemy terrorists worldwide still fight with the AK-47 assault rifle.

Experienced American combat soldiers learned to respect the AK-47 assault rifle. It was such an effective battle rifle that one sniper bunker mate of mine went to the field carrying a captured AK-47 along with his standard sniper rifle. He told me that his M-14 sniper rifle was great for long distances, but if the enemy were closing in, he would rely on his AK-47 to beat back enemy assaults. He used the Russian rifle knowing that he risked being shot at by other GIs, who might fire at his position because of the sound of the foreign weapon. He said he would take that risk because more than likely the AK-47 would save his life.

Our main battle weapon in Vietnam was the M-16 assault rifle. The Army issued me my own personal M-16 automatic rifle. It was a gas-operated lightweight rifle made of plastic-like composite material and stamped aluminum. It fired 5.56-mm rounds in either a semi-automatic or full automatic mode, depending upon the setting of the select fire switch. When it was first issued to GIs in Vietnam, the M-16 caused much controversy because soldiers discovered, to their horror, that the rifle would jam easily when exposed to dirt or mud. It is reported that many Marines were found dead with their M-16s stripped down because the Marine was trying to clear his jammed rifle. In contrast, the AK-47 almost never jams, no matter how bad the environmental conditions, because the tolerances between its parts are not as tight as tolerances in our military rifles. The AK-47 literally has space for the dirt or mud to escape. I do not have the military expertise to fully evaluate the assets and liabilities of the M-16. Many rifle experts love the weapon and its present day successors, especially after the weapon was modified with a chromed chamber to prevent fouling. Yet it is safe to assert that most rifle experts would agree that our M-16 was outclassed in the Vietnam War by our enemy's AK-47.

My roommate in Vietnam was a sniper. The weapon he was issued was a specially machined M-14 sniper rifle with a noise suppressor attached to the end of its barrel and a large Starlight night vision telescope as the weapon's optics. The Starlight was an early night vision device; it magnified the night's ambient light from the stars and the moon, thus enabling the sniper to see figures and movement in the darkness.

We stored both weapons horizontally, one over the other, on hooks attached to one of our cubicle/room plywood walls. Many a night I would sit on my bunk and stare at the two rifles hanging on our wall. I was bemused by the contrast between the two rifles. When placed next to the sniper M-14, my M-16 looked like a toy. It was plastic and stamped metal, while the M-14 was made of wood and specially machined steel. The M-14 was big and heavy; my M-16 was much smaller and lighter. The M-14 fired specially machined 7.62-mm sniper rounds; my M-16 fired the lightweight 5.56-mm round. The M-14 looked and felt like a genuine battle implement. Attaching an Army bayonet to my M-16 made it look like a joke.

Yet when I fired my M-16 at the rifle range on full automatic, I was always impressed. My M-16 was remarkably well balanced and lightweight; I could hold and fire it with one hand, yet its rate of fire was that of a true machine gun. When firing in full automatic mode, I could empty a ten-round clip in just a couple of seconds. As long as I kept it clean, my M-16 was a capable weapon.

Dusters

Our battalion always kept our firebase's perimeter tightly closed for troop protection. Whenever any units or personnel visited, they always stayed within our berm. There was, however, one exception to this basic rule, when an artillery-tracked vehicle rotated to our firebase.

In the American Army there is a standard allotment of guns and troops for each artillery brigade. One section of an artillery brigade is devoted to anti-aircraft defense. During the Vietnam era that meant two self-propelled tracked vehicles on which were mounted twin rapid-firing 40-mm anti-aircraft cannons. They were nicknamed "Dusters" and were designed to attack enemy aircraft in a conventional war. But Vietnam was not a conventional war; yet they were still standard issue for every artillery brigade. Since in South Vietnam neither the VC nor the NVA fought with any type of aircraft, these anti-aircraft artillery (commonly called "Triple A") vehicles had to be used in a non-conventional manner. Early in the Vietnam War our Army commanders discovered that, when their barrels were lowered, these 40-mm cannons were extremely effective in an anti-personnel role. By 1969 the two Triple A tracked vehicles were separated. Each track would routinely rotate to the various firebases of the brigade to aid in protecting the base from the enemy's ground attacks.

During their regular base rotation schedule, one of these Dusters traveled to our firebase. I was impressed by their audacity as they drove themselves in the early morning down the dirt road to our front gate. They stopped the Duster at the gate and stayed there. That night, when our gate was closing, I expected the Duster to move inside our perimeter. They did not. Their crew possessed so much confidence in the destructive lethality of their twin 40-mm cannons that they attempted to provoke the VC into attacking. They positioned the Duster right outside our front gate, daring the VC to attack. However, the VC must have respected the Duster's firepower, because they did not bite on the bait. There was no attack that night. The next day the Duster drove on to the next firebase. In my memory that was the one and only time they protected our firebase.

Spooky

One night an unknown number of VC fighters attacked our Tan Tru firebase. It was early in the evening, so I had not yet reported for duty at the TOC. Instead, I was resting in my bunker when the enemy's AK-47 bullets started hitting our firebase. The alarm siren screamed out. As usual, our mortars quickly responded with defensive fire. Within a few minutes I could hear the artillery rounds that had been fired from other firebases in our defense hitting outside our berm. But after a short time, the artillery support stopped. I became worried, because this cease-fire was highly unusual and dangerous. Wondering what was happening, I stepped outside of our bunker and looked up at the night sky.

I spotted a slowing flying airplane firing down around our perimeter. Spooky was on the scene. "Spooky" was the nickname of a World War II vintage transport plane, the Douglas AC-47, which is the military version of the famous DC-3 transport, that had been weaponized by adding three multiple-barreled M134 mini guns. The M134 mini gun was a multiple-barreled machine gun that was capable of firing 2,000 to 6,000 7.62-mm rounds a minute. The mini gun owes its design to the American Civil War's Gatling gun. The problem with machine gun rapid fire is that in a short time the barrel overheats and warps, causing loss of accuracy at first, and later possibly a barrel explosion. Dr. Richard Gatling solved the problem by designing a multiple-barreled weapon. Each barrel only shot every sixth bullet, which prevented overheating. The barrels on his gun, the first machine gun, were rotated by hand with a crank. The General Electric Company adapted Dr. Gatling's design by creating a six-barreled machine gun whose barrels are rotated at high speed by an electric motor.

The Air Force converted the two-engine AC-47 transport plane into an air-to-ground attack fighter by custom fitting three M-134 mini guns into the left side of the fuselage, the pilot's side. One gun fired out of each of the two windows on the plane's side, and the last gun fired out of the loading door. Spooky accomplished its ground attack missions by executing what is known as a pylon turn over the enemy's position. That means the plane would fly one big left turn, circling the enemy while firing all three of the mini guns at him. Spooky could rain fire down on the communists over a longer period of time than any conventional fighter plane's strafing runs.

18. Weapons

Spooky's three mini guns added a massive amount of firepower to any battle. Spooky normally carried 54,000 7.62-mm rounds on a mission. Each of the mini guns would usually fire about 4,000 rounds a minute. The M134 mini gun is capable of placing an explosive bullet in each square yard of a football field in three seconds. Its guns fired so much ammunition that Air Force crews were issued a snow shovel so that they could shovel out the open door the thousands of shell casings accumulating inside the plane.

Every fifth round on Spooky's ammunition belts was a tracer, which is a bullet that burns bright red as it approaches the target. That night I watched the mini guns' many tracers combine into a ruby-red glowing stream of fire that looked like a science fiction death ray. What I heard was not like any rapid-fire machine gun; I could not distinguish the sound of individual rounds. Instead I heard a BBBRRRAAATTT that sounded like an evil high-speed chain saw. It was truly an awe-inspiring sight and sound. It made me feel completely secure, because I knew that the VC ground attack could not survive that volume of firepower. Later I learned that I had been correct in my conclusion. The Air Force has documented that, during the Vietnam War, no American base or outpost was ever overrun when defended by AC-47 gunships.

After they were deployed, the VC soon learned to fear these gun ships. Because of the red fire spewing out of the plane's guns, the VC called them dragon ships, like mythical fire-breathing dragons. The VC's name for the gunship led American soldiers, who heard the term, to start calling the plane "Puff the Magic Dragon," later shortened to "Puff," from the popular folk song of the era sung by Peter, Paul and Mary. So the AC-47 carried two names: Puff and Spooky. Either way it was one badass weapon system.

Spooky gunships proved so successful that after the war our Air Force expanded and updated the concept of a ground attack aircraft by heavily arming the C-130 Hercules transport airplane. The Hercules is a four-engine propeller plane that is converted to the AC-130 Specter by installing 40-mm and 25-mm rapid fire cannons as well as sophisticated sensors and computer fire control systems. What is most interesting to an artilleryman like me is that the Air Force added a 105-mm howitzer to the plane's armament. Their engineers had to develop a huge shock absorber to dampen the cannon's massive recoil or the howitzer would have shaken the plane apart. Today the AC-130 Specter is an even more lethal killer than the Spooky gunship was in Vietnam. The AC-130 Specter gunships were successfully deployed in the Gulf War and are still being used effectively in the War on Terror.

Rocket–Propelled Grenades

The NVA and the VC also were also well equipped with Russian rocket-propelled grenades (RPG), either the RPG-3 or later the heavier RPG-7. The RPG-7 is a shoulder-fired weapon that fires rockets equipped with an exploding warhead. They were originally designed as anti-tank weapons; however, in Vietnam the communists utilized them in anti-personnel, anti-machine gun and anti-aircraft roles. RPG-7s have been used to shoot down even our newest combat helicopters. They were state-of-the-art in the Vietnam era, and our enemies in the Middle East are still today, 50 years later, killing American soldiers with RPG-7s.

19

Atypical Duty

Usually my nighttime duty at the TOC was routine. I would receive, plot and clear all the potential targets for the next day's airmobile assaults. Then I would establish communications with our battalion's listening posts and ambush patrols. Finally, I would establish contact and constantly monitor the radio traffic from our infantry companies out in the field. On many nights nothing happened. However, there were always nights when something unusual would occur. The following vignettes describe a few of the non-routine nights.

The Prisoner

It was the dry season in Vietnam's Mekong Delta, which meant sweltering heat with no rain. On November 20 I was on duty at the TOC during the day because my counterpart, a liaison sergeant, was in the field functioning as the RTO for the battalion's FO. I was monitoring a battle; the fighting had been savage and brutal. Our battalion's infantry had suffered casualties, but the grunts were killing the VC. One infantry company had even captured a prisoner, which was a rare and valuable accomplishment.

The Viet Cong prisoner had been badly wounded; he was shot in the face. Yet our medics had kept him alive, and he was being flown back to our firebase to receive further medical treatment. I heard over the radio that the Huey had landed and the prisoner was being treated on the main dirt road in our firebase. Once I heard the news I decided to check him out.

I ran down the road to the group of soldiers who had gathered around the prisoner. It was the first and only time I actually laid eyes on our enemy up close and alive. The VC prisoner was sitting cross-legged and stoic on the side of the road. He was not a young man, but I always found it difficult to estimate the true age of the Vietnamese; their harsh lives aged them prematurely. He was wearing peasant garb, not black pajamas but rather a dirty grey pajama-like outfit. The most striking aspect of his appearance was his face. The bottom half of his face was covered by a large Army bandage. He had no jaw; it had been shot off. He was bleeding profusely.

What struck me was that he was conscious, sitting there Buddha like, staring straight ahead, showing no evidence of pain. He may have been given a morphine shot, but I do not think any American could have sat so still without their jaw, no matter how much morphine they had been given. On the mean streets of Pittsburgh I had seen street toughs screaming and writhing in pain from a broken jaw. I have trouble even imagining how much agony a man suffers by having his entire jaw shot off. This was our enemy, enduring immense pain. I was impressed by his high pain tolerance. I recall thinking, "No wonder we are not winning the war against these people; they can out-suffer us."

The scene soon turned into a surrealistically absurd situation. I heard the colonel yelling over a radio at the young lieutenant who had been ordered to interrogate the prisoner. The colonel was cussing out the lieutenant for not obtaining the intelligence information the colonel wanted. The lieutenant tried to explain to the colonel why he was failing: "The prisoner was not talking because he had no jaw." To the colonel, crazed with anger, this reality did not matter; he wanted that information. The lieutenant was becoming helplessly frustrated. I turned and started walking back to the TOC, feeling sad and disillusioned about the war.

Mercenaries

During South Vietnam's wet season from May to October, we all had to suffer through frequent monsoon rainstorms. These were terrible downpours when the rain would be so heavy that it would hurt. When I was caught in a monsoon, it felt like I was standing under a raging waterfall. The monsoons were so heavy that even the VC and NVA did not conduct raids or ambushes when one hit. The American Army tended to follow their example; it was as if the war had stopped.

Consequently, I was curious when one night at the TOC I received a radio message from one of our night outpost guards (our battalion always had soldiers manning listening posts outside our perimeter, no matter what the weather) reporting that he had spotted a strange company of soldiers. He had spotted approximately 70 military men walking down the middle of the dirt road that led to our front gate. What was most surprising was that under closer inspection this mystery unit appeared to be led by an American officer wearing standard American jungle fatigues, but displaying no unit patches.

The mysterious group was allowed to walk up to our gate and the night duty officer granted them permission to enter. They marched straight to our TOC. Their commander came in to talk privately with our night duty lieutenant. I was not privy to the discussion; the matter was obviously top secret. I was so intrigued by the visit that I had to study these courageous strangers, who marched down the middle of the road at night inviting an enemy attack. What I discovered peering out of the TOC entrance was a group of tall Asian men wearing different colored military-style fatigues, standing in formation under the monsoon downpour. Interestingly, each man was armed with a different type of machine gun. I recognized a number of Russian and Chinese machine guns that were usually crew-served weapons, yet here one man was carrying them. I did not know many of the other foreign weapons, but every one looked formidable and intimidating.

All of the men were standing, not at attention, but rather in a relaxed upright stance, somewhat like our Army's parade rest posture. I was impressed with how stoic and quiet they were. They took no notice of the heavy downpour. None of the men were talking. They simply stood there under the downpour appearing lethally sinister. They reminded me of the Caribbean pirates of old, but much more menacing. They remain the most intimidating group of men I have ever laid my eyes on. The night duty officer told me later that they were Chinese mercenaries, guns for hire that were obviously fighting for us.

Their captain came out of the TOC, took his place at the head of the formation and marched his Chinese mercenaries through our firebase and out the back gate, never to be seen or heard of again. Seeing these clandestine warriors unnerved me; I had been exposed to an alternative evil universe of money for blood that I had not known really existed. It was another one of my surreal "Wizard of Oz" moments in Vietnam: "Toto, we are not in Kansas anymore."

Black Ponies

As I reported, our battalion, the 2nd Battalion of the 60th Regiment of the 9th Infantry Division, worked closely with the Navy in Vietnam's Mekong Delta. The Delta was extensively pervaded by rivers, canals and small waterways. The Brown Water Navy patrolled all of these waterways.

One night I was monitoring our radios for communications with our companies and outposts in the field. Our battalion was not working with the Navy this particular night; yet in the middle of the night, I heard this voice on our radio network:

> "Firebase Four, Firebase Four, this is Black Pony, Black Pony. We are looking for trouble. Got any?"

That last sentence and the follow-up question have to be the coolest message I ever heard in Vietnam. "We are looking for trouble." I loved its courageous bravado. These men had brass cojones.

The Black Ponies were a Navy fighter wing made up of Douglas A-1 Skyraiders, World War II fighter planes that were heavily armed with machine guns, rockets and bombs. They were propeller driven and relatively slow compared to jet fighters, but in Vietnam that was an advantage. They could fly lower and slower during their attacks on enemy positions, dropping their bombs and firing their rockers with greater accuracy than jets, which flew so fast that their accuracy was compromised. These older fighter planes with their hefty firepower were more lethal in ground attacks than jet fighters, and thus were feared by the VC. The sortie that night consisted of two fighters patrolling along the Mekong Delta's rivers and canals.

I clearly remember my answer. It was the most enjoyable communication I ever had with air power during the war:

> "Black Ponies, Black Ponies, this is Firebase Four.
> Negative, negative. No trouble"

20

Unique Nights

NVA Officer

One night as I was reading in my bunker cubicle, one of my bunker mates came running through the narrow hallway that ran through the center of our bunker, shouting out, "Gooks inside the wire … gooks inside the wire." I jumped out into the hallway to gather more information. Apparently he was walking on our metal pathways when he spotted an enemy soldier off in the distance, crouched over and running through our firebase. This VC or NVA soldier had somehow avoided our guards and crawled over the concertina wire barrier that created the major part of our perimeter defenses. Concertina wire is a type of barbed wire formed into large coils that can be expanded like a concertina to install. Along its wire coils are razorlike steel strips designed to snag and rip open anyone trying to penetrate the barrier. Concertina wire and metal stakes were used to build military obstacles. We had multiple layers of concertina wire completely surrounding our firebase.

Obviously, this was an extremely dangerous situation. Even just one enemy soldier inside our defenses could wreak havoc. He had the potential to kill or wound many of our men, who were relaxing or sleeping, unaware of the threat. And there might possibly have been more than one enemy loose inside our firebase.

After the alert, men were locking and loading their M-16 rifles all across the firebase as they prepared to join in the hunt for the intruder. However, I knew that many of these hunters had been smoking pot or drinking alcohol heavily prior to the alert. To me, there were few situations more precarious than having drunk or stoned GIs wandering around the firebase ready to shoot their loaded M-16 automatic rifles. Considering that it was a dark night, shots could easily be fired at anything suspicious that moved, including other Americans. Consequently, I decided to lock and load my M-16 rifle, but not to join the unorganized hunt. I retreated to a corner of my cubicle/room, sat on the floor with my lights out and awaited developments.

After about an hour of futile searching, my bunker mates returned to report that the elusive enemy had avoided capture. Happily, none of our men had been hurt during the hunt. Still, everyone remained on alert. I imagine no one slept soundly for the rest of that night; I know I did not.

About eight weeks later, an NVA officer surrendered to one of our infantry patrols. During his interrogation, he claimed that he was the one who had penetrated our barbed wire defenses that night of alarm to gather intelligence. At first our intelligence officers refused to believe him. How could he have gotten through the impenetrable concertina wire? The prisoner offered to demonstrate how he had crawled through our concertina wire barrier.

I was on duty at the TOC and could not attend the demonstration, but I learned about it from a friend who was there. First of all, he reported that this NVA prisoner was a short sinewy man. My friend said that it was hard to believe what he observed even though he saw it happening right before his eyes. The prisoner was able to crawl into the coiled wire on his fingertips and toes and then slowly contort his body in ways that a normal man could not. He was agile and flexible enough to extend and bend his arms, legs and trunk in unusual postures that enabled him to inch through the coiled wire without getting caught on any of the numerous razor shards welded to the wire coils.

I wish I had been there to observe his circuslike act. If I had been present, I would be able to recount his contortions more descriptively. I still have trouble imagining how someone could perform such a feat. Yet this NVA prisoner did in fact slowly work his way through the barbed wire barrier. I was told that the intelligence officers were incredulous. Previously, they had all thought, as I had, that it would be impossible for a human being to crawl through thickly coiled concertina wire. They had trouble believing their eyes. But here was evidence that they had been wrong.

However, there was one positive outcome from this NVA officer's unique demonstration. Our firebase commander ordered that the combat engineers attach various alarms to the concertina wire. The engineers filled empty C-ration tin cans with small pebbles and then hung dozens and dozens of these cans from various points along the concertina wire. If anyone simply touched or shook the barbed wire, the tin cans would rattle out an alarm to our guards. They also strung out trip wires inside the coiled barbed wire attached to deadly Claymore antipersonnel mines. If a Viet

20. Unique Nights

Concertina barbed wire barrier on dirt road behind our firebase that led to ARVN artillery firebase.

Cong tripped one of those wires, he would trigger an explosion of 700 small steel balls that would chop him down like a gigantic shotgun shell blast. Because of the surrender of this NVA officer, our firebase's concertina wire perimeter defenses became even more treacherous and difficult to infiltrate.

The Replacement

In the first years of the war the American military shipped intact military units over to Vietnam. These were infantry battalions and artillery batteries that had trained together, sometimes for years, and then were shipped to Vietnam together as a group. These soldiers knew each other and had bonded together. However, as the war dragged on and our casualties mounted, more and more individual soldiers were sent to Vietnam to replace those lost in battle. This individual replacement strategy was the same strategy used by our military during the final years of World War II. Postwar studies found that these new replacements suffered the highest casualty rates of that world war. In the Vietnam War, a replacements or "newbie" would simply be plugged into an existing unit. The most obvious problem with the individual replacement strategy is that the soldiers in the company receiving the replacement did not know the new guy.

The following vignette that occurred at our firebase during my Vietnam War tour is a tragic example of the consequences of plugging new replacements into existing army units.

One of the saddest duties a soldier could be ordered to perform in Vietnam was to pack the personal belongings of a comrade who had been killed in the war. The commanding officer would usually order friends of the dead man to sort through his personal belongings and censor them before packing them up to be shipped home to his family. This was done to prevent any unsavory material that might cause the soldier's family distress from reaching them. For example, all pornographic material that the dead man might have possessed would be destroyed. Also, any repugnant photographs of dead or desecrated enemy bodies would be censored. For example, pictures of beheaded VC or body parts were burned. If the friends of the dead man found anything close to these disagreeable examples, it was their duty to destroy it. The soldiers I talked with considered this duty highly distasteful, but at the same time recognized the necessity of the task.

However, there was one time when the squad mates of one of our battalion's soldiers killed in action did not have to censor and pack his belongings. This is how it happened. One day during the dry season an infantry replacement, who had just completed his initial weeklong acclimation training at Dong Tam, reported in to one of our battalion's company commanders. The captain then assigned him to one of his four maneuver platoons, which meant that the new guy was taken to his platoon's living quarters bunker, shown his new bunk and introduced to his platoon's leader and also to the platoon sergeant. The platoon sergeant then ordered the new guy to join a squad of infantrymen who were to go out on listening post duty that night.

The poor fellow only had enough time to throw his duffel bag on his bunk, grab some chow at the mess hall and prepare his weapon and gear for that night's patrol duty. When darkness fell, he reported to the squad leader, who marched the six-man patrol out through a designated opening in our concertina wire perimeter and over the rice paddies to the patrol's night position. After pulling his turn at guard duty, the new guy fell asleep. While he was sleeping, his squad leader decided to move the squad's position for better security. But the squad leader was so stressed that he forgot about the newbie. He moved the rest of the squad to a new location without waking the new guy or taking a head count.

Later the new guy awakened and to his dismay discovered that his squad had left him. He was alone and lost on his first night in the forbidding rice paddies of the Mekong Delta. He panicked and started running through the rice paddies, searching for his lost squad. He did not realize how dangerous it was to move through the Delta at night with a rifle in his hands. Tragically, an ARVN night ambush patrol did not recognize him as a panicked American soldier; the ARVNs shot and killed him.

The next morning, after the poor fellow's body was found, everyone at our firebase was talking about the squad leader's screw up. I did not blame the ARVN patrol; I blamed the squad leader's failure of leadership as the reason for the new guy's needless death. All of the soldiers whom I talked with agreed with me. Military leadership's basic requirement is to keep track of all your men. I never learned how that squad leader was punished, or if he was punished at all. To me he deserved a severe punishment.

The irony of the story is that the men in the dead man's squad did not have to go through his personal belongings to pack them up. The unfortunate new replacement *never unpacked*. His squad mates simply grabbed his full duffel bag off of his bunk and threw it on a truck that delivered it to a shipping port.

Alien Worlds

Things are quiet at the firebase on the night of July 20, 1969. I am alone in my bunker's cubicle/room, listening to Armed Forces Radio on the small transistor radio my fiancée gave me. The Army's radio station is broadcasting live the flight of America's Apollo 11 rocket to the moon. I listen through those last tense moments as the landing module finally touches down on the moon. A few minutes later, Neil Armstrong climbs down the module's ladder and becomes the first human to ever step on any solar body other than Earth. Once on the moon, he broadcasts his famous statement: "That is one small step for a man, one giant leap for mankind."

As I sit there processing the historic event, a question runs through my mind: "They're on the moon, I'm in Vietnam. Why can't everyone just stay home?"

21

Drugs

Today it is widely known that many American soldiers used various mind-altering substances—drugs—during the Vietnam War. Apologists have claimed that after all it was the 1960s and that our soldiers' drug use was not any more prevalent that the drug abuse occurring on America's college campuses at that time. But I seriously doubt that college students had easy access to such hard-core drugs as heroin or opium, which were readily available in South Vietnam. My estimate of American soldiers' drug use during my year in South Vietnam is as follows: approximately 60 to 70 percent of the soldiers used or abused alcohol. It was the most widely available mind-altering substance in South Vietnam. Every American base had an officers, "an NCOs" and an enlisted men's club, which were basically bars serving beer and liquor at discount prices. When the mess hall's menu was grilled steak, the cooks would also serve beer with the dinner.

Approximately 20 to 25 percent of our soldiers used marijuana, nicknamed pot or weed. In Vietnam, marijuana was widely available and easily obtained. In fact, it grew wild in the Mekong Delta. I never paid a single cent for marijuana in Vietnam. Our headquarters medic would routinely go out on Med-Caps, which were humane patrols to Vietnamese villages in order to provide free medical care to the Vietnamese civilians. On Med-Cap patrols my bunker mate medic would have the children collect the wild marijuana while he was treating their elders. This medic would return to our bunker carrying large bags full of marijuana, which he would share with his bunker mates.

Five to ten percent of our battalion's soldiers abused illicit opiates and amphetamines. Vietnam is located near the Golden Triangle of Thailand, Cambodia and Laos, the largest region of opium cultivation in the world. In Saigon, opium dens were nearly as common as neighborhood bars are in our country. Some Vietnamese would snicker at the American soldiers for smoking pot, because they preferred to abuse the more intoxicating opium. Liquid amphetamines in small glass vials were also available. I was told that most of these vials were stolen from NVA medical kits and then sold on the black market. These hard drugs were not as easily obtained as marijuana, but were available for the right price. It seemed that in our battalion the group with the highest rate of heroin or opium use was the infantry.

A small percentage of American soldiers abstained from all mind-altering substances. Another small percentage of soldiers was high on the war. They relished the adrenaline their bodies produced in combat; we called them "war junkies" or "excitement junkies." They wondered, "Why do you mess around with that shit, boy? Let's go out and kill some Viet Cong." My response was, "Take a chill pill, dude."

I was introduced to marijuana while stationed at my artillery brigade's base camp in Tan An. Back on my college campus in Pennsylvania I had known of a few students who indulged in marijuana use, but they were only artsy types who were not close friends of mine. Never really fond of alcohol, I was curious about the intoxicating effects of marijuana, but I had never had the opportunity to experiment with it in the civilian world. During my first weeks in country at Tan An, I worked with a number of soldiers who smoked marijuana regularly. They offered to "turn me on," drug slang for introducing me to the drug.

The first time I smoked a marijuana cigarette—a joint—was on someone's bunk in a tent full of pot smokers. I did not feel much difference; I was probably anxious concerning the effects and resisted. My more experienced "trip director" friends assured me that my reaction was a common one among first time users. They encouraged me to try again.

The next night I returned to their tent to smoke another joint. A couple of minutes after inhaling the smoke, a transcendent wave of euphoria swept over me. My senses were more acute. Everything seemed more enjoyable. The rock and roll music playing sounded better. Colored lights seemed to glow in brilliance. I was physically relaxed but mentally stimulated. For the first time since my arrival in South Vietnam, I felt relatively calm. About one-half hour later I experienced an intense appetite, what pot users called "the munchies." Food tasted better. Eating a couple of candy bars felt like I was indulging in the most mouthwatering gourmet feast ever.

Later, at my permanent assignment to the firebase near Tan Tru village, I lived with the other soldiers of the battalion's Headquarters Company in a bunker/barracks. Most of the men who lived in that bunker regularly smoked marijuana. Soon I was a member of our bunker's pothead majority.

The unofficial leader of our group was Doc, the Med-Cap medic, who was from California. He told me that he had made a vow to himself that if he had to spend a year in Vietnam, he would be high on some drug the entire year. As far as I could tell he stayed true to his vow; he was high all the time.

My living quarters bunker had a small wooden deck off to one side of one of the entrances. One end of the deck led to our cold-water shower, and the other end formed a porch where the pot users would gather. I became a regular attendee of the nightly gatherings. The marijuana was potent and abundant. On some nights there would be ten men smoking numerous joints on our small porch. Many nights a cloud of pot smoke would form and hover over our porch. We consumed our drug of choice in the open; no one ever harassed or attempted to stop us.

Ever since my early adolescence I had consumed alcohol, but I had never developed dependence on the drug. I never enjoyed being excessively intoxicated; drinking to the point of becoming sick enough to vomit was not appealing. And I hated the next day's alcohol hangover. I drank mainly to be part of a group, such as my high school football team, who would meet at a team member's house to drink a keg of beer. They would drink until they were sick. I would drink a little, not enough to get sick. Once I tried marijuana I was pleased to discover the absence of residuals. I never became sick from smoking pot, and there were no hangovers the next morning. I delighted in marijuana's relaxing effects. In my Vietnam War, marijuana became an escape, allowing me to take a chemical vacation from the hell that engulfed me. It was a medication for the chronic depression that I felt from being exposed to so much death.

At first we all smoked hand-rolled marijuana cigarettes, but smoking these joints was harsh on our throats. Then one of my bunker mates brought back a water pipe he had bought in Saigon. Using this device that filters the smoke through a liquid made the pot smoke much milder on our throats. Water pipes soon became our standard method of smoking pot. I bought one from a Vietnamese vendor and started experimenting with various liquids to replace the water. After trying a number of substances, including various types of liquor, I found that root beer soda in the water pipe brought a subtle, enjoyable sweetness to the smoke.

Another one of the guys brought a large, six-hosed hookah back from his R&R leave in Bangkok, Thailand. The hookah's clay bowl was as big as a medium-sized flowerpot. One evening six of us lay around the owner's cubicle-sized bunker room smoking from the hookah. We each had our own long hose and mouthpiece, enabling us to literally lie back and indulge.

21. Drugs

The hookah's water bowl was a large one, filtering so well that the smoke we inhaled was quite mild. It was a truly luxurious way to smoke marijuana. At one point, after we had smoked a bowl, the owner emptied out the ashes and commented that we needed more weed to refill the bowl. Each man reached into the thigh pocket of his jungle fatigues to pull out an ounce bag of marijuana. We all stopped to look at each of us holding out his own large plastic bag of weed and burst out laughing at our absurd level of overabundance.

One possible attempt at limiting our marijuana use was the construction of a Military Police (MP) bunker near our deck. Before this event we did not have MPs stationed at our base. I am still not sure what their specific job was on our small firebase. Yet all of a sudden there they were. But they were never a problem for us marijuana smokers. The MPs never raided our smoking deck. It may have had something to do with the fact that one night one of our snipers, for some unknown reason, took his special M-14 sniper rifle into the MP barrack/bunker and shot off a round down the bunker's hallway. Fortunately no one was hurt, but I believe that this crazy behavior frightened the MPs. Another reason for their acceptance of pot was that they had no shower in their bunker. The engineer who had built ours gave the MPs permission to use our shower. Many times they had to walk through the cloud of smoke to reach the shower, which they did without any of them saying a word about the marijuana cloud.

Eventually, I became a frequent marijuana user. Much later I realized that after smoking marijuana, I never dreamt or experienced any terror-ridden nightmares. Today I am sure that this side effect unconsciously contributed to my being a marijuana smoker while in Vietnam.

I used marijuana not only for the intoxicating bliss, but also because it identified me as a member of a rebellious group of soldiers. Most of the soldiers at our firebase were either "juicers" who indulged in alcohol or "heads" who indulged in marijuana. The juicers included the vast majority of officers and non-commissioned officers, who frequented the officer and NCO clubs to imbibe the cheap liquor. To us heads, the juicers were the establishment that we looked down upon. We considered ourselves the hip elite compared to the conservative, old fashioned alcoholics. We were the rebellious future. I felt personally close to the men who shared my rejection of the military mindset and the war, even though they were from vastly different backgrounds. Most of the men I smoked marijuana with were from small rural southern towns or from big city California.

Our bunker's deck was not the only center for smoking pot at our firebase. Occasionally I would wander over to the artillery section of our firebase to smoke with the men who actually aimed and fired the big guns. A few of us would climb up to the roof of a sandbagged artillery bunker to share some pot. From our elevated vantage point we could see for miles over the flat Mekong Delta. It was the perfect place to watch the war. On some nights we could see other American or ARVN bases under attack. The white flares, the VC green tracers and our helicopter gunship's red tracers always made a vivid light show.

At other times we could watch the Navy gunboats patrolling the nearby rivers, shooting up anything that moved. One time, a patrol of two Navy swift boats was cruising the river when a VC sniper fired one shot at the boats. In response, both boats turned toward the shoreline and opened up with their twin Browning 50-caliber machine guns and their direct-fire mortars. The Browning 50-caliber is a formidable weapon, firing six-inch-long rounds that were designed to destroy military vehicles. They were devastating against personnel. Both of the Navy gunboats were firing vast amounts of ammunition; their red tracers looked like fireballs pounding the jungle. Their heavy machine gun rounds were literally chopping down trees, while their mortars were blowing up any possible sniper hides.

What happened next made us Army types smirk. After both gunboats had expended all their ammunition, over a thousand rounds, they raced off to their homeport, reloaded and returned at speed to the site of the sniper fire. Again they opened up with their heavy machine guns and mortar against a sniper that we were sure was long gone by then. The only thing the sailors were accomplishing was punishing the land where the sniper had hidden. We Army types were jealous, in one sense, because the Navy did not have the same type of fire discipline necessary in the Army. When an infantryman is carrying all his ammunition on his back, he is forced to conserve his rounds in the fear of running out. In contrast, the Navy did not, as was demonstrated by their overreaction to one single sniper round that night.

My favorite memories of the artillery bunker roof smokers were the ones when we shared our favorite pastimes from back in the world. Six to eight of us would gather in a circle on the bunker's roof. As we passed around a number of joints, one man would teach the rest of us about what made his favorite way of spending time back in the world so special. The job of the storyteller was to describe in vivid and precise details what made his hobby so enjoyable. For example, one guy described how much love and work he put into modifying his stock car in order to race it on dirt tracks over the weekends. Another man shared his love for raising hunting dogs and then hunting game with their help. Our job as the audience was

21. Drugs

to put ourselves in his place as best we could to understand the storyteller's reasoning. We all empathized with him as to why that activity was so much fun. We reinforced the storyteller with our comments and questions. There was one rule: No criticisms or putdowns of the man's hobby. If you could not empathize, then you just kept your mouth shut.

It was mind-expanding fun, and being totally stoned helped. It enabled me to enter the storyteller's worldview and to understand men from other parts of the country. Now I do not recall exactly what my contribution was when it was my time to share my favorite back-home activity. It could have been going to our county's North Park for our large extended family picnics and swimming in the park's enormous pool, whose width was equal to a regulation Olympic pool's length. Or perhaps it was playing sandlot touch football with my suburban neighborhood friends. I am sure that I did not share my true favorite pastime, which was and is reading, both fiction, including the classics, and non-fiction, especially psychology. I did not want to be considered a bookworm egghead. No matter what pastime I did share, those lonely soldiers smoking marijuana with me on that bunker roof encouraged me to travel back mentally to a happier time, to escape from the 24-hour, seven-day-a-week red alert stress that I felt trying to survive in Vietnam's Mekong Delta war zone.

Towards the end of 1969, a replacement second lieutenant was assigned to our firebase. He was short and overweight, and he stood out in his new deep-olive-green-colored jungle fatigues. The color betrayed him as a soldier just beginning his Vietnam tour. After many trips to the Vietnamese laundries, that peculiar olive-green fatigue color was washed out into the brownish green colored fatigues of a "short-timer," a soldier who had a short time left in his Vietnam tour. This particular lieutenant was driven by a desire to quickly make a name for himself on the firebase. My theory is that he was an ROTC graduate who needed to impress our commanding officer, the colonel.

Our firebase's perimeter was ringed by a dozen or so guard bunkers manned by the infantry. Every night one of these bunkers was designated as the site of a marijuana smoking party for the infantry potheads. They would spread the word on the day's location informally as they passed each other the firebase's raised walkways: "Bunker ten, bunker ten." One night soon after he arrived, the newbie second lieutenant crawled up to one of these guard bunker parties. The low crawling lieutenant was fortunate that he was not misidentified and shot by one of the infantry guards. I am not sure what exactly he said to the smokers, but I was told that while the lieutenant talked, the potheads stuffed their bags of marijuana between the bunker's ammunition boxes. No one was arrested because the lieutenant was unable to find any marijuana on any of the men; however, he did confiscate a large cache of marijuana. He brought the confiscated pot to the TOC and insisted that the night-duty officer store the pot in the brigade's large safe located in the colonel's office. The night-duty officer complied.

On duty that night at the TOC I witnessed the new lieutenant's marijuana drama unfold. For the rest of the night I wondered what the colonel's reaction to the pot cache would be. Up to that point the colonel did not seem to be concerned over the marijuana smoking that occurred on his firebase. I was worried that this new lieutenant's actions might trigger a crackdown on pot use. The newbie lieutenant showed up bright and early the next morning wearing his new green fatigues to claim his confiscated marijuana from the safe. The colonel appeared at the TOC about eight in the morning.

Once he saw the colonel, the lieutenant went up and briskly saluted the commander. He explained how he had confiscated the marijuana, which he proudly displayed to the colonel. He went on to explain how this particular batch of marijuana was an extremely potent variety. Slowly, the colonel, with a slight smile on his face, looked up and said, "Lieutenant you don't expect my men to smoke anything but the best." I almost fell off of my chair in surprise over the colonel's statement. The lieutenant's prideful spirit burst like a punctured balloon; he turned and scurried out of the TOC, leaving the pot behind.

A couple of weeks later I overheard the colonel talking to another officer in the TOC. He commented that the marijuana smokers were not like the alcoholics, who regularly fought with each other, sometimes even pulling out weapons. Evidently, the colonel tolerated the marijuana use on the firebase because the marijuana smokers were passively nonviolent. Another important consideration was that all his men followed the cardinal rule that there would be no substance abuse, neither alcohol nor marijuana, out in the field on combat patrols. Everyone knew the stories about American soldiers who were found at their guard duty posts in the field with their throats cut and the marijuana joint still dangling from their mouths. Decompressing with your chemical of choice was acceptable in the relatively safe confines of the firebase, but both enlisted men and officers did not tolerate substance abuse in the field.

Another drug that was widely abused at our firebase was amphetamine, nicknamed speed. The type of speed that was most available was methamphetamine, a potent but toxic central nervous system stimulant. It came in liquid form inside small glass vials. The background story I was told was that these vials were a standard component of the NVA medical kits and were utilized for the drug's anti-fatigue properties. Supposedly the drug also had some analgesic effects. Most American abusers would break open the tip of the vial and pour the liquid speed into a soft drink; Coca Cola was the favorite brand. Then the drug user would drink his "atomic highball." Once his body had absorbed the drug, the speed freak would be extremely energized for 48 hours. He would lose his appetite and be unable to sleep, yet feel focused. He would posses the grandiose sense that he could do anything he put in his mind to. After two days, the amphetamine would finally be metabolized from his body, triggering a collapse of fatigue. Most abusers would sleep through the entire third day.

Once, in the middle of my tour, I experimented by mixing and drinking an atomic highball. The subsequent effect was far from a pleasant experience. The energy grew and grew inside of me to an anxiety-provoking peak. I was wired to the max, literally trembling in my skin. That night I reported to my duty station at the TOC as usual, but I could not sit still. I paced and paced around the TOC trying to burn off the energy. When the intensity became too much to tolerate, I asked the duty officer for a break, went to my bunk and rolled a few joints. I quickly smoked one outside of the TOC. It calmed me down enough that I could sit at my duty station and monitor my liaison radios. Unfortunately, after a time the pot wore off, allowing the amphetamine-induced frenzy to return. Luckily, it was a slow night without any enemy contacts. Five times that night I had to slip out of the TOC to smoke another calming joint. I suffered intense agitation throughout the rest of my duty shift, but at least I did finish the shift.

Of course, I was not able to sleep the next day; instead I smoked a ton of pot to slow down. I did not eat at all that day; instead I drank a dozen sodas. Overall my feeling during the long hours of my amphetamine experiment was that of terribly intense turmoil. I had been tense enough from dealing with the war; I really did not need this chemically induced mania on top of the anxiety that already existed. I vowed never to ingest another amphetamine again. To this day I have been true to that vow.

However, different men react differently to the same drug. Our firebase commander, the colonel, had a major who was his executive officer, that is his direct assistant, the officer second in command of the entire firebase. This major developed a reputation for being a gung-ho John Wayne type of soldier known for his courage. Whenever there was any type of contact with the communists at our firebase, he would run towards the action to join the fight. He even had a 50-caliber heavy machine gun bolted onto the back of one of our standard jeeps. When there was any attack on our perimeter, he would drive his armed jeep to the section of the perimeter under attack to add his heavy machine gun's firepower to our perimeter's defenses. For weeks on end he always seemed to be up and involved while I was on night duty at the TOC. If any of our infantry companies in the field or one of our listening posts came under attack, he would take over the radios and command our soldiers' response to the attack. Usually during these attacks he would order me to provide artillery clearances for the various specific targets he picked, which I would do ASAP.

During calm nights I would see him sitting at the colonel's desk poring over maps as he conceived and planned various new combat operations to win the war. I am no military strategist, but the plans that he did explain to me seemed dangerously dubious. For example, one of his plans was to have two- or three-man dog teams patrol along the Cambodian border to act as an early warning system for any communist cross-border attacks. Thankfully, these perilous operations were never implemented.

At other times the major would come to the artillery side of the TOC to talk with me. Even though I had no desire to communicate with this officer, when a major decides to engage in friendly conversation with an E-4 Liaison Specialist, the specialist has to listen. I provided the major a living ear at which to direct his overflowing speech. He must have enjoyed the experience, because he started to share personal information with me. One night he was describing his family back home in the world, including his three children. To provide a visual aid, he dug out from his wallet a family picture taken just before he was deployed to Vietnam and showed it to me. What struck me at once about the picture was that all of the people in it, including the major himself, were obese. There he was in the photo, a smiling fat father with his obese family. In stark contrast, the major standing beside me was so thin that he looked almost anorexic. I switched my gaze back and forth a few times until the light bulb in my mind clicked on: "Oh wow, the major is a speed freak."

It turned out that I was correct. Later, I learned that the major was buying his methamphetamines from the young Vietnamese woman who ran the checkpoint store on the corner of Highway 4 and our dirt road. She was a well known drug pusher. His methamphetamine abuse explained his staying awake at night in the TOC devising new war plans. After being awake all night, the major would return to his quarters, shower, and don a freshly starched uniform. When the colonel arrived at the TOC about eight in the morning, the major would approach him all bright eyed and snap the colonel a good morning salute. He would act like this for two days; then he would disappear for a day as he collapsed from the speed wearing off.

After I returned to the world from Vietnam, I corresponded with a few of my friends who were still serving at the firebase. One friend wrote me that this major had been busted by the battalion's new commanding officer, a no-nonsense colonel whose prior assignment had been at the Pentagon. The colonel somehow discovered the major's methamphetamine addiction, relieved the major of his command, and had him flown back to the States in disgrace. This is another example of the surreal nature of America's Vietnam War, where senior officers were serious drug addicts.

Many Vietnamese sneered at American soldiers who indulged in such a soft drug as marijuana because their illicit drug of choice was opium. Since Vietnam was close to a world-class source of opium, the Golden Triangle, uncut opium, as well as one of its synthesized products, heroin, was readily available. In Saigon, opium dens were as numerous as beer pubs in Milwaukee. Basically, I avoided opium because I was afraid of heroin. As an adolescent in inner city Pittsburgh, I had known a neighborhood teenager who had died from a heroin overdose. The police found his body on a toilet with the hypodermic needle still stuck in his arm. He died because the heroin he injected was not diluted with other substances and thus was lethally potent. The fact that he died because his dealer had been honest with him by selling him uncut pure heroin was too weird for me. I decided to avoid that world of hard drugs where honesty could kill.

However, I did smoke a little opium once in the form of an opiated joint. Vietnamese dealers would empty the tobacco out of a pack of regular cigarettes and then refill the paper sleeves with marijuana. Next they dipped the joint in liquid opium, producing what was known as an OJ, an opiated joint. The dealers repacked all 20 cigarettes in each pack, resealed them, and then sold them to American soldiers.

During my last month in Vietnam I was lucky enough to receive orders for a one-week R&R in Hong Kong. On the first day of my leave, hitchhiking down Highway 4 to Saigon, I stopped at an outdoor Vietnamese restaurant for lunch. I sat alone at a picnic table in the restaurant's large backyard, which was filled with other American soldiers enjoying the fare. At a nearby table I spotted a group of American soldiers staring at me with suspiciousness. After a few moments I realized that they wanted to smoke a joint but were worried about my reaction. I signaled to them to go ahead, I did not mind. Relieved, they smiled and lit up. When they were leaving one of the guys came over and handed me a pack of the pre-rolled marijuana cigarettes. I thanked him and smoked one for dessert.

Soon I was back at the highway, thumbing for a ride to Saigon. The solo driver of a 12-ton Army supply truck stopped to pick me up. This 12-ton designation is not the weight of the vehicle, but rather it is the size of the cargo load it can carry. It was an immense truck; I had to climb up to the cab's passenger seat, which was at least 15 feet above the road. The friendly driver agreed to take me all the way to the Tan Su Nut Airport. Once he started to drive, I quickly realized that he was probably on some type of amphetamine, because he drove like a maniac. He was driving 60 to 70 miles per hour down the highway that was crowded with Vietnamese civilians riding their ubiquitous motor scooters. My driver never slowed down when he came up upon a mass of small scooters; instead, he blew his deafening steam horn and plowed ahead. The motor scooters had to move to the side or risk being run over. The mass of scooters opened up a gap as we barreled straight ahead into the center of the mass.

By this time, the effects of the joint I had smoked had started to kick in. I felt extremely intoxicated; in fact, I started to semi-hallucinate. In my mind, the truck was a whale I was riding on and the motor scooters were schools of thousands of small fish that had to swim out of the way or be eaten by my whale. It was the wildest ride of my life, the truck driver speeding and me imagining that I was riding on a whale, while he continually hit the earsplitting horn. Many of the small scooter/fish just barely got out of our way. I came to the realization that the joint I had smoked must have been an OJ coated with opium. I am still not sure that the 12-ton truck did not run over one of those Vietnamese motorbikes; it would have felt like just another bump in the road.

That night at Tan Su Nut Airbase I slept in a holding barracks while waiting for my flight to Hong Kong the next day. Before we boarded our jet we had to listen to a lecture about the Chinese port city, which included a stern warning about China's strict laws against illegal drugs and the harsh punishments for breaking their drug laws. Believing the message, I reluctantly threw the rest of my pack of OJs into an overflowing amnesty barrel.

A percentage of Vietnam's opium was processed into heroin that was sold to the Americans. I was told that it was pure enough that soldiers did not have to inject the narcotic into their veins; instead they could snort it like cocaine to achieve the desired rush. Yet some soldiers still preferred to inject the drug. Most of the hard-core heroin abuse at our firebase—and I suspect the rest of South Vietnam at the time—was done by the infantry. The infantryman in the Vietnam War was the soldier with the most dangerous duties, the soldier most likely to die. As Kris Kristofferson wrote,

21. Drugs

"Freedom's just another word for nothin' left to lose." Unfortunately, a growing percentage of infantrymen in the war felt that they had nothing left to lose. To them, heroin did not seem that treacherous.

At our firebase in 1969 my estimate was that less than five percent of our battalion's infantrymen were abusing heroin regularly. In 1970, after the Nixon excursion into Cambodia, an army friend wrote to inform me that our infantry battalion was one of the first to cross the border into Cambodia. He went on to complain that the NVA in Cambodia opened the heroin floodgates; it was everywhere. He reported that in our battalion, in Cambodia and afterwards, the percentage of frontline infantrymen abusing heroin rose to 30 percent. Heroin abuse became such a widespread problem for our Army that late in 1970 MACV commanders initiated a program of narcotics drug testing. Every American soldier returning to the world had to pass a narcotics screening test before he would be allowed to fly home.

Back at our firebase, the time of the week that alternative drug use became most apparent was eight o'clock on Sunday nights. That was the time that the "Sergeant Pepper's Radio Show" would be broadcast over the Armed Forces Radio network out of Saigon. The name came from the highly popular Beatles album "Sergeant Pepper's Lonely Hearts Club Band." The show's disc jockey called himself Sergeant Pepper. He devoted his one-hour radio show to the latest rock and roll music. It was the only show on Armed Forces Radio that broadcast this style of modern music, which was wildly popular among the young soldiers. As I walked around the firebase during that eight o'clock hour, just about every young soldier had his transistor radio tuned in to Sergeant Pepper's show. Even though I did not carry my radio with me, I could still hear the hard-core rock music that permeated the atmosphere. Rock music filled the air.

To prepare for Sergeant Pepper's show, every pothead at the firebase would smoke a bowl or a joint before the broadcast. It was on one of these Sunday nights, when I was strolling near an infantry company's area, that I saw an infantrymen dancing by himself, holding his radio on his shoulder near his ear. What upset me was that the young soldier still had the hypodermic needle stuck in his arm. He was so zoned out on heroin that he did not feel the needle in his arm. His lack of awareness reminded me of my dead neighborhood buddy back in Pittsburgh; to me, here was even more evidence that the potent narcotic heroin was a drug to be feared and avoided.

22

Officers

In my job as a liaison specialist in Vietnam, I was required to work closely with the captains who were the artillery forward observers for the commanding officer of our firebase, the colonel. When we were in the field, I was the RTO for the FO captains. Interestingly, three out of the four FOs that I worked with during my yearlong combat tour were West Point graduates. They were considered an elite subgroup among all officers in the Army. My impression was that other officers, who had won their commissions through other pathways like ROTC or OCS, were a bit jealous of the military academy graduates. Officers sometimes dismissed them as "ring knockers" because of their annoying habit of knocking their West Point rings on tables during meetings. They felt that all the West Pointers were on the fast track for promotions; they were probably right.

Working closely with these military academy graduates for many months allowed me to know them quite well. I found them all to be intelligent individuals, but their knowledge base was limited to military technical matters like the ranges of the various artillery pieces or artillery tactics. It seemed to me that their knowledge base and military skills had few applications within the civilian world. I never heard any of them ever discussing literature, philosophy or even politics. What they did discuss most of the time was their primary concern, their careers. They would even talk to me about their careers. I recall one captain telling me why he chose West Point over the Air Force's military academy and his plans for military promotion—like I cared. Of course I listened anyhow, because initially officers intimidated me. But the West Pointers annoyed me, because they would discuss such personal matters with me as an equal when we were both on night duty at the TOC, but at all other times they would adopt an attitude of elitism and rank.

Generally I found the West Pointers' thinking to be rigidly formal

22. Officers

and conventional. Their lack of creative thinking in the all-important matter of military strategy and tactics is demonstrated by their use of our six 155-mm self-propelled howitzers. None of the academy graduates ever utilized these powerful tank-like howitzers' ability to move themselves into positions where their fire could be more effective in battle. They never ordered the tracked howitzers to leave our firebase. In contrast, the one captain FO that I worked with who was not an academy graduate—he had earned his commission through Officer Candidate School—ordered two of our 155-mm tracks to drive into the field and then moved them into positions so that they could fire directly into the enemy's bunkers to destroy them (see Chapter 23 for details of this operation). It was such a successful tactic that day that I never understood why the West Point officers did not use it.

But at least all West Point captains were intelligent. That was not always the case with other officers. One night I reported for duty at the TOC to find a captain at my post holding a small notebook of mine. He braced me, which means he ordered me to stand at attention while he harassed me. He asked me angrily if it was my notebook. It was one those medium-sized secretary notebooks that had lined pages that you could flip over the top to reach another page. I answered in the affirmative. On slow nights at the TOC it could be difficult for me to stay awake, so to stimulate myself I challenged my mind by working out chemical equations. Back in college my choice of a major was between chemistry and psychology. Of course I chose psychology, but I always kept an interest in chemistry. In order to keep myself alert on those slow nights, I drew chemical symbols like benzene rings and wrote out various chemical equations and worked out their mathematical solutions in that notebook. These equations were what the angry captain found written out in that notebook.

I do not recall exactly what he said as he screamed into my face, mainly because I was confused. I did not understand, at first, why he was so upset. I knew that I had done nothing wrong; other soldiers, even officers, were regularly writing letters home or reading paperback books during slow nights at the TOC. These activities were always better than falling asleep on duty. After a minute or two of his shouting venom, I finally understood the source of his anger. He was yelling at me for solving chemical equations because it was an intellectual exercise that he could not do himself. My educational skills had activated his deep-seated sense of inadequacy. This officer simply could not handle the possibility that there was an enlisted man who might be better educated or more intelligent than he. It was ironic to me, because this particular man was our firebase's S-3 Intelligence

Officer. No wonder our battalion's intelligence reports were always so erroneous. After a few more minutes of yelling, he threw the notebook on my makeshift desk and stormed out of the building.

My respect for him diminished because of this episode. It was not that elevated beforehand anyway. Few enlisted men respected him after one mortar attack on our firebase a number of weeks earlier. While the alarm siren screamed, he ran to a bunker, but stubbed his toe while running on the metal walkway near the entrance. A few days later the enlisted clerk, who had to type up the recommendation, told anyone who would listen that this captain had written up a recommendation for himself to be awarded the Purple Heart Medal because of his stubbed toe during the mortar attack. The pathetic ending to the story is that he did receive the medal, which was really meant to award soldiers who suffered the pain of serious battle wounds. It was not meant for stubbed toes caused by hiding from enemy mortars. But as enlisted men frequently complained, "Rank has its privileges."

It seemed to me that most Army officers possessed a vanity concerning their rank, yet for some the arrogance was more directly apparent. A vignette about such arrogance follows the required background information.

In Vietnam there were soldiers who worked as journalists writing biographical portraits of soldiers in the war, intended for the soldier's hometown newspaper. We had one of these journalist sergeants visit our firebase for a couple of weeks. During his stay he interviewed me, along with a number of other guys, and then wrote up a short article about my work in the Vietnam War. He sent my article in to his superiors, thinking it would be sent to Pittsburgh papers. Instead, for whatever reason, the *Stars and Stripes*, which is a daily newspaper that is published by the American military, picked up the article. On Sunday June 22, 1969, the *Stars and Stripes* newspaper published this complimentary article about me and distributed it to our military across South Vietnam and across the world. The following are a few sentences from the short article:

> The ace up the sleeve of every infantryman in Vietnam is the awesome firepower of artillery. A few words spoken over a field telephone can, in a matter of minutes, bring high explosive shells screaming through the air to rain havoc on an unsuspecting enemy. Along with the enormous capabilities of artillery goes a tremendous amount of responsibility. Spec.5 Walter F. McDermott, Jr., an artillery liaison specialist from Shaler Township, Pa., is one man who shoulders much of that responsibility. [*The author then goes on to describe my job, especially that of firing at the doughnut hole in encirclement operations. He concludes:*] A cool head and a quick mind are vital assets of an artillery liaison specialist. McDermott and men like him hold the lives of our fighting men in their hands.

22. Officers

Both the journalist who wrote the article and I were surprised by the newspaper's publication. He was proud that one of his articles was published in this prestigious newspaper with its widespread distribution. I was pleased with the positive nature of the article, which I had not read beforehand.

A few days later, I was on day duty at the TOC when a major whom I had never met called me on our landline telephone from somewhere in South Vietnam. He yelled out a question that was worded something like: How did a lowly enlisted sergeant like you—the journalist had given me a fictitious promotion in the article—get such a positive article written about him in the *Stars and Stripes*, while an exalted officer like himself, a major, did not have one written about him? I had no good answer for this major; I pleaded that it was not my fault. After venting all his displeasure at the perceived unfairness of the newspaper and at me, he hung up. I was astonished by the call. The egotism of these types of vain narcissistic officers appeared limitless. These negative encounters with our officers prevented me from ever developing a firm respect for our officer class.

23

Vietnamese Enemies vs. Allies

During America's War in South Vietnam, we were forced to fight two different types of communist forces. There were the Viet Cong who were a guerrilla force made up of the rank and file fighters and a leadership cadre. Many in the ranks were rice farmers or shopkeepers during the day and guerrillas fighters at night. Some VC were forced to fight by the communists, who would come into a village and conscript the military-aged Vietnamese men. If any of the men or their families resisted, the communists would execute the man's family in front of the rest of the village as a lesson. As one might predict with these types of troops, VC units were not always hard core. Sometimes, when outnumbered and overwhelmed by our firepower, they would surrender. They did not wear any specific uniform; instead, they usually wore the standard garment of the rural Vietnamese farmer, black pajamas. The VC were primarily armed with the ubiquitous AK-47 assault rifle. Their heaviest weapons were rocket-propelled grenades, light machine guns and small mortars. Occasionally they would capture a larger weapon, mainly from South Vietnamese Army forces, the ARVN, like the recoilless rifle they captured and fired at our base.

In sharp contrast was the other communist group we fought against, the North Vietnamese Army (NVA) forces. They were trained soldiers, they wore uniforms, and they possessed heavier weapons than the VC. For example, they routinely utilized crew-served .51-caliber heavy machine guns. Their most unnerving asset was their total dedication to the communist cause. They were willing to die for their beliefs. I read that before a NVA soldier marched from North Vietnam down the Ho Chi Minh trail to the war in South Vietnam, his family would hold a funeral service for him. How eerie that must have been for the young soldier, conscious at his own funeral, knowing that his family expected him to die. The ceremony must have changed the soldier's heart to its core. Now he was expected to die; perhaps now he felt that he must die in order not to disappoint his family.

One battle in our AO during my tour exemplifies the NVA soldier's dedication to a communist victory in the war. Most of our battalion's fighting was against the VC, but occasionally one of our infantry companies clashed with an NVA unit patrolling in the South. The unique feature of this particular battle was that our new battalion FO, the only one of the four captains I worked with who was not a West Point graduate, actually had two of our self-propelled 155-mm howitzers drive into the field in support of our maneuvering infantry companies. He was the only one of those FOs who ordered this aggressive maneuver.

This particular day, our infantry company discovered and fought against an NVA patrol of 18 soldiers. Reeling from the initial firefight, the outnumbered NVA patrol retreated into prepared concrete bunkers hidden in the bush. The Vietnamese had been fighting on this land for decades, first the Japanese, then the French and now the Americans. During those long years of war, they had built secret bunkers into the jungle and rice paddies. These bunkers were expertly camouflaged, well hidden from anyone not privy to prior knowledge of their location. Housed in their six concrete bunkers, that small NVA patrol may have felt safe from the attacking American infantry.

Nevertheless, they had not anticipated the artillery support our infantry had on call that day. I monitored the firefight by listening to the artillery radios back at our TOC. My FO, flying with the battalion's colonel in the C&C helicopter, saw the retreating NVA flee into the bunkers. He ordered the two 155-mm self-propelled howitzers that had been standing by to move up into a position that would enable the big guns to fire directly into the enemy's bunkers. These concrete bunkers were so well built that they were impervious to small arms, mortars and most indirect fire artillery; but direct fire at close range from a powerful 155-mm howitzer is an entirely different matter. Our self-propelled guns came up on line and fired two 155-mm rounds into the nearest bunker. The explosive shells blasted the first bunker, killing all the NVA soldiers hiding inside. Rather then firing immediately at the next bunker, the colonel magnanimously ordered up a psychological warfare helicopter. Once on station, it circled the bunkers while a South Vietnamese soldier using a loudspeaker tried to talk the rest of the NVA patrol into surrendering. In response, the trapped NVA soldiers tried to shoot down the circling psychological warfare helicopter.

So, rather reluctantly, the colonel ordered the 155-mm howitzers to fire on the next enemy-occupied bunker. Two 155-mm artillery rounds destroyed that second concrete bunker. Again the psychological helicopter was ordered up, and again the South Vietnamese psychological war specialist pointed out to the remaining enemy the hopelessness of their position. He promised the NVA that they would receive humane treatment if they would surrender. The trapped NVA fired on the psychological warfare helicopter a second time. In response, our artillery FO ordered the howitzers to fire into the third occupied bunker. Again and again the same tragic scenario repeated itself until there was only one intact bunker left.

This time the South Vietnamese psychological warfare specialist changed his argument. This time he emphasized to the last remaining NVA soldiers that all their comrades were dead and now no one would know that they had surrendered. Yet again as their answer the remaining trapped NVA soldiers fired upon the psychological warfare helicopter. And again the battalion's FO ordered the 155-mm howitzers to destroy the last enemy bunker, which they did. In total, six sturdy concrete bunkers were destroyed, killing all the NVA soldiers inside. The next day I talked with an infantry sergeant who had inspected the bunkers after the fight. He reported that there were no real bodies inside; instead there were only body parts and bloodstains left. This was the type of NVA soldier we were fighting against; they fought to the death.

The antithesis of our NVA enemies were the soldiers of the nation America was sacrificing so much for, the soldiers of the Army of the Republic of Vietnam (the ARVN). Now I am sure that there were some courageous South Vietnamese soldiers and some aggressive ARVN outfits. For example, I read that one South Vietnamese Marine division fought courageously defending Saigon during the last days of the Vietnamese Republic in 1975. However, there were no aggressive ARVN units in our division's area of operations during my 12-month combat tour.

To me, there was ample evidence that the ARVNs were an ineffective fighting force. For example, my roommate was a sniper who shot a specialized M-14 rifle equipped with a state-of-the-art starlight telescope. The starlight was a night vision device that enabled him to see figures at night by amplifying any ambient light produced by the stars and the moon. He and his sniper partner were assigned to accompany ARVN night patrols because our top commanders did not issue starlight telescopes to the ARVN. Our leadership feared that they would be too easily captured from the South Vietnamese soldiers, thus eliminating one of our major military advantages. My roommate told me that, when he was patrolling with an

ARVN unit, he would stay up on guard all night because the ARVN guards would routinely fall asleep.

Another example of ARVN military ineffectiveness occurred when a small team of VC sappers heavily damaged the ARVN artillery firebase located about one and one-half miles from our firebase. Educated Americans know that during World War II the desperate Japanese ordered volunteer suicide pilots, the kamikaze, to dive their bomb—laden airplanes into America's fighting warships. They proved to be devastating weapons, the first intelligent guided missiles, sinking five warships and heavily damaging 23 others during the 1944 battle of Leyte Gulf. Yet few American civilians know that the Vietnamese communists also fielded suicide soldiers, who were called sappers. The enemy's sappers were suicide squads who would attack a defended position armed with nothing more than a satchel filled with military-grade explosives. Fighting against an enemy that is willing to die is an intimidating challenge. These are people who do not possess our Western culture's value of the sacredness of life; instead they glorify death. Their disregard for their lives makes them formidable opponents, allowing them to concentrate solely on destruction.

A team of only six sappers attacked the nearby ARVN artillery firebase. One communist sapper threw himself on top of the ARVN barbed wire defenses while detonating the explosives he carried. By blowing himself up, the suicidal sapper created a breach in the South Vietnamese defensive perimeter that the other members of his suicide team exploited. They ran through the firebase throwing their satchels at their primary targets. These five sappers destroyed two 105-mm howitzers; they destroyed the ARVN tactical operations center, killing the commander and the other officers inside. They also killed a dozen South Vietnamese soldiers. Of course, the sappers were all eventually killed, which they knew would happen, yet they were able to inflict damage out of proportion to their numbers. The VC sappers were able to close in on the ARVN's barbed wire defensive perimeter because the South Vietnamese had not posted any night guards or sent out any listening post patrols. Most of the firebase's soldiers had been asleep when the attack began. Any responsible army would have posted night guards at least.

Yet, in spite of such evidence, in July 1969 President Nixon and Secretary of State Henry Kissinger initiated their "Peace with Honor" plan to end America's involvement in the Vietnam War. It was a two-pronged plan; the first prong was the Vietnamization of the war. This meant turning over combat operations to the South Vietnamese Republic's Army after a massive resupply of weapons and enhanced combat training of their army. The second prong was the gradual withdrawal of American combat troops from Vietnam, starting immediately with the first 25,000 men. Two first rate divisions were selected for this first phase of the withdrawal plans: the 3rd Marine Division and my own 9th Infantry Division.

The news excited the men at our firebase, which was abuzz with multiple rumors because the actual details of our pullout had not been finalized. One of the rumors claimed that the division would be deactivated and the division's soldiers transferred to other active divisions. This was an unpopular option for most troopers. A second rumor was that the division would be withdrawn to Hawaii, where it had previously been stationed before being sent to Vietnam. This rumor was greeted with excitement from all the enlisted men, especially among the married soldiers. The division's reassignment to Hawaii would mean that the wives of married soldiers could join them there.

Most of the enlisted men at our firebase were borderline irrational in their excitement over the imminent possibility of being withdrawn from Vietnam and transferred to Hawaii. It seemed to be the only topic of conversation among the men for weeks. Every married man reported that he had shared the possibility with his wife, who became as excited as he was. What was not clear was how many of the men had told their wives that was only a rumor and not an established plan. One of my married bunker mates became so convinced that the Hawaii move was going to occur that he had his wife pull up stakes and move to Hawaii in anticipation of his arrival.

I was not as excited about the Hawaii rumor as were most of the other men. My main military goal was to keep my active duty time down to the absolute minimum. I did not want to return to the world only to be reassigned to another Army base, even if it were in Hawaii. I would still be in the Army and subject to the spit and shine harassment that happens at all Army bases that are not in a war zone. For me, one of the advantages of my assignment to Vietnam was that I was not subjected to that kind of harassment; for example, our bunkers were never inspected, nor had I stood in any formation during my entire tour at the Tan Tru firebase. At that point, I knew I could not tolerate a Stateside sergeant screaming at the top of his lungs into my face over some minor infraction. I was truly afraid that I would lose my temper and punch the offensive sergeant.

Also, reassignment would mean a longer stay for me in the active duty Army. I was already looking forward to the standard early out for Vietnam War returnees who had less than five months left in their enlistment. The Army had discovered that short-term assignments after Vietnam service did not work out well. Too many Vietnam returnees were involved in multiple disciplinary infractions at their new bases. The new Army policy was to discharge returning Nam vets with less than five months left. A move to Hawaii would mean that I would be required to serve out all the time left in my two-year enlistment contract. Plus, through sad experiences I had learned not to trust the rumor mills. Who really knew what was going to happen to the division once it was withdrawn from Vietnam?

After a few weeks, the details of the division's Vietnam withdrawal were finalized. Two of the 9th Division's brigades would be withdrawn. The third brigade would be administratively assigned to the 25th Division, stationed at Chu Chi. All of the long timers were reassigned to the third brigade, meaning that most of the men to be sent home were already scheduled to DEROS soon anyway. I had to sign a waiver to stay with the division in Vietnam. For me that waiver meant more time in Vietnam, but less total time in the Army. I would not have signed if I had been an infantryman, but I had a relatively safe job. As for the 9th Division, it was not sent back to Hawaii but rather deactivated. The trooper whose wife was awaiting him in Hawaii was reassigned to Alaska. That was the price he paid for allowing his emotions to overwhelm his rational thinking to the extent that he believed in positive Army rumors.

Little changed in our day-to-day military life at the firebase after the reassignment of our brigade to the 25th Division Infantry. It was only when a commander of an ARVN infantry battalion moved onto the firebase three months later, in October 1969, that things changed. As part of his Vietnamization program, President Nixon had ordered more joint operations, which meant that the ARVN and the U.S. military were to plan and execute combat missions together. As a result of Nixon's orders, the command group of an ARVN infantry battalion moved onto our firebase. Our TOC was literally divided in two, with the ARVN colonel and his staff manning the other half. The two commanders were to meet daily to plan missions.

Our battalion did go on joint missions with the ARVN battalion, but the frequency of these major operations decreased. The ARVN colonel did not have the heart for combat. Whenever our commander would offer a realistic combat mission plan, the ARVN colonel would always present some excuse why the plan would not work. If our intelligence reported that the VC were operating in one area, the ARVN colonel would insist that combat patrols were necessary in another area. Our colonel became increasingly frustrated and angry at the ARVN colonel's reluctance to fight the communists.

I knew that Vietnamization would not be successful because I saw how the ARVN had become too dependent on the Americans to do the fighting for them. Yet I never voiced my concern, not that it would have changed the reality, because I was happy that President Nixon was starting to bring American soldiers home from that ghastly war. In a real sense the military supplies we gave to the South Vietnamese ended up arming the NVA. One crazy incident during the end of my tour exemplified this sad trend. A large group of South Vietnamese from one region joined the ARVN army together. American and ARVN drill instructors provided them military training at a base near Saigon. During this basic training they were issued M-16 automatic rifles and taught how to shoot them. Upon graduation, the entire company marched off of the field with their rifles, never to return. They had deserted en masse. We had trained and armed an entire VC company of over 100 men.

Fortunately for me, this partition occurred when I had only three months left in my Vietnam tour, because it proved to be an unworkable alliance. None of our soldiers I talked to or worked with had any confidence in the South Vietnamese Army's fighting spirit. I developed a pessimism that the Vietnam War was not going to end in victory. I feared that once we left Vietnam, the Republic of South Vietnam would collapse because its Army could not stop the communists' military forces. In fact, that is what happened. When the Vietnam War's final battles were fought in 1975, too many ARVN soldiers simply abandoned their weapons and operational military vehicles, including tanks, helicopters and jet fighters, and ran from the attacking NVA. Many discarded their army uniforms and changed into civilian clothes to hide within the masses of refugees. The military supplies and weapons we gave the South Vietnamese Army ended up making the NVA one of the best equipped armies in the world in 1975 and for the decade afterward.

To this day I do not understand the stark differences between the North and South Vietnamese armies' fighting spirit. They were all one people, were they not? Without going into a complete history and political analysis, it is a fact that the Republic of South Vietnam was not a democracy. The "Republic" was in reality a corrupt oligarchy where rich families maneuvered politically to continue their dominance. The ARVN army officers were appointed, not by merit, but by their political and economic connections. Many ARVN officers would not turn in accurate casualty counts of their soldiers because ARVN soldiers were paid through their officers. Senior officers would receive the pay of dead soldiers and keep it. The consequence of this corruption was that Army units that appeared to be at a certain strength level on paper were in reality much weaker. The communists

23. Vietnamese Enemies vs. Allies

were able to field soldiers who were fully dedicated to a communist victory, while the ARVN soldier was not dedicated to his government's cause. The ARVN soldiers I met were only ready to fight until the last American soldier died.

There is a human aside to Vietnamization and our firebase's joint operation plan with the ARVN. During the switchover, two ARVN interpreters were assigned to live in our headquarters bunker. They shared one of our bunker's cubicle rooms and tended to keep to themselves, which was fine with us. However, there was one problem: the Vietnamese soldiers liked to listen to their favorite Vietnamese music at night loudly. It could be heard throughout most of our bunker. Their "music" was irritatingly horrible. It is much like their language, atonal and unmelodic. To us it sounded like screeching cats, and it had the same effect on our nervous systems as the sound of someone scratching fingernails down a chalkboard.

We had to fight back, so a couple of my bunker mates, who had bought high quality stereo sound systems while on R&R, set up their large speakers right next to the plywood dividers that defined the ARVN cubicle room. That night when the ARVN played their music, my mates played Detroit Motown rock and roll music at top volume. Aretha Franklin proved to be the best at drowning out the invasive Vietnamese music, especially her song *Respect*. It did not take but a few nights of our musical bombardment for the ARVN translators to get the message to turn down their radios. After a time we turned them into American rock and roll fans.

24

Saigon

One of the few regrets I have about my Vietnam War tour was that I did not spend more time in Saigon. I visited the capital enough to be seduced by its primeval energy. It was an exotic, savage and exciting city. First of all, the traffic was insane. The streets were six lanes wide, but most of the time the Vietnamese did not bother to follow lanes. Only a few of Saigon's streets had traffic police at intersections; on every other street it was every man for himself. The blare of horns was constant. I learned to drive my jeep with one hand always on the horn and the other on the steering wheel.

I never knew what I would see next while driving in Saigon. The first few times I drove in the capital I had to avoid large numbers of armored cars, military supply trucks, bicycles, horses, cars, motor scooters, an elephant and the stray tank. The most interesting vehicles were what I will call motor scooter trucks and buses. The front of these vehicles was the standard front tire and motor of a Vespa-like scooter, but the back would be an oversized metal box with either seats (the bus) or cargo (the truck). Most were overloaded, with some passengers barely hanging on or goods of all types piled much too high.

Adding to the city's menace was a Vietnamese woman, obviously a communist sympathizer, who, dressed in a traditional Ao Di outfit, drove her motor scooter all over Saigon. The Ao Di is an ankle length dress with a slit up both sides to the waist; usually it is worn over white pants. If she spotted an American officer standing on a city corner, she would pull up, take out her pistol and shoot the unlucky officer at close range. She would then drive off, to disappear into the crowds of motor scooters that flooded the roads. American officers in and around Saigon all called her the "Dragon Lady." She targeted only officers; she had not attempted to assassinate an American enlisted man during my time in Vietnam. Her targeting

24. Saigon

Typical Saigon street scene.

officers was a harsh reality for them, but another one of the many benefits of my decision not to become an officer.

One time I had to hitchhike to Saigon to clear up some paperwork problems with my R&R application. After solving the problem, I decided to walk to the major American Military PX located in the Cholon area—the Chinese sector of Saigon. As I walked Saigon's streets protected only by my M-16 rifle and flak jacket, I never felt totally comfortable. I was continually scanning rooftops for snipers, while carrying my rifle at the ready position. Yet I saw American soldiers walking around unarmed. I saw American civilian couples holding hands like they were on a date. It was a strange new world to me, as if the war did not exist in Saigon.

What did exist was the black market. The closer I came to the PX, the more dishonest Vietnamese civilians would approach me to buy American dollars. It seemed that every corner had illegal currency exchange deals occurring. Black market traders would try to buy dollars with military payment certificates. I refused all these illegal advances because I knew those greenback dollars would later be used to buy guns and ammunition to kill Americans. Those crooked money dealers were like flies; as soon as I would rebuke one, two or three new ones would approach me. When I refused, they would inevitably cuss me out in their broken English.

The central PX was a large department-like store that sold everything from food, liquor and cigarettes to small appliances, including refrigerators and televisions. However, all the goods were rationed to prevent them from being resold on the black market for inflated prices. Every new American soldier was issued one ration book filled with tickets for various items. That soldier could buy only one television set or one stereo system, because there was only one ticket for each appliance in his ration book. There were dozens of Vietnamese black marketers loitering on the steps leading to the PX, each one trying to buy ration book tickets from me or attempting to pay me to use my tickets to buy an item for them. Again and again I refused their solicitations, making me unpopular with this unsavory group.

I pushed my way though all these black marketers to enter the store. After turning in my ration ticket for the item, I succeeded in buying an oscillating electric table fan. I kept it for the rest of my tour; it became one of my prized possessions. It was the only weapon I had to battle the intense temperatures of the Mekong Delta, where the heat and humidity at high noon could make a strong man faint. Well built and reliable, the fan cooled me when sleeping through the sweltering 100-degree days. At night I would unplug my fan and carry it to my duty station at the TOC. There were many days when it blew cooling air on me for the full 24 hours.

Another time I was alone in Saigon exploring the streets of a wealthy residential neighborhood near the Saigon River. Many of the houses were two or three story villas, most of which were secured by high walls surrounding the perimeters of the property. Yet down the same residential streets there would be an empty lot filled with piles of rubbish and rubble between two villas. As I walked near the villas, a putrid odor permeated the atmosphere. The closer I came to the river, the stronger the stench. I soon discovered that the source of the foul odor was the garbage-filled Saigon River. Here in an elite Saigon neighborhood, the river functioned as the city sewer. The resulting vile odor was nauseating. I did not understand how the wealthy residents tolerated it.

In early May I was asked by an officer to drive him and two of his officer friends to Saigon. The reason for the request was that Army regulations insisted that officers were not allowed to drive jeeps themselves. Only enlisted men were authorized to drive. I agreed, even though it meant that I would be losing sleep because I still would have to man my post that night at the TOC. There were no days off in my Vietnam War.

The day came and there I was driving three young officers to Saigon. My preferred route would have been along the highway, but the officers insisted that we take a more direct route over back roads. This was a dangerous route, along dirt roads that sometimes went directly through small villages, where the inhabitants viewed us with what appeared to be hostility. I was driving slowly and so near to them that I imagined how easy it would be for one of them to throw a grenade or fire a magazine of AK-47 rounds into our jeep. It was only a 15-mile drive to Saigon, but it seemed like it took hours to complete on those narrow dirt roads.

The officers directed me to the city's business center, specifically to one of the quality Vietnamese restaurants in that area. We entered one that was dark, clean and air conditioned. The establishment impressed me as calm and civilized, an oasis of peace in a war zone. The staff was pleased to see us. When we first sat down, a servant boy brought us steaming perfumed towels to wash the road dirt off of our hands and faces. I ordered Chinese mushroom soup, egg foo young, jumbo shrimp wrapped in pork, and ice cold fruit for dessert. We all shared huge bowls of steamed rice with pork bits and ice-cold clams. The restaurant only served cold beer in gallon size containers. Needless to say, the officers and I drank our fill. We all relished the exceptional food and pitched in to pay the bill of 36 dollars.

Afterward, the officers explained to me that they were going to drop me off and they were going to go on with the jeep to party on their own. Somehow, having dinner with an enlisted man was acceptable, but having him accompany them on their bawdy partying was unacceptable. They had me drive to an ARVN base somewhere in Saigon, and there they dropped me off at the ARVN's officers' club. Obviously it was acceptable for an American enlisted man to drink and socialize with ARVN officers, just not American ones. I do not know why they picked that particular place for me. They had me dismount the jeep and told me that they would return to pick me up in three hours. Since I did not know what to expect, I did not argue against their plan. Anyway, I had no choice.

When I walked through the doors of the ARVN officers' club after driving in the tropical sun, I was initially blinded by the darkness. As my eyes adjusted to the dim light, I saw that the club was packed with both American and ARVN soldiers dressed in their fatigue uniforms. I noticed a ruckus at the far end of the bar. A group of not-so-young Vietnamese bar women were arguing and wrestling with one another. After a minute their brawl ended in true Darwinian fashion, as the strongest and fiercest female fighter emerged the victor. I was her prize.

She broke free of the pack, grabbed me by the arm, and pulled me into a booth in the back room. After the shock of the entrance wore off, I ordered a beer and a drink for her, which I later learned was another

Vietnamese scam, Saigon tea. I paid a stiff price for an alcoholic drink, yet she would be served only lukewarm tea in a cocktail glass. This bar girl would then receive a share of the profits generated by her drinking the inexpensive tea. As soon as the waitress left, this bar fighter grabbed my crotch. Surprised, I tried to push her hand away, but, as noted, she was strong. By now my eyes had adjusted to the dim light, enabling me to see that all around the room were American soldiers, mainly Air Force types by their uniforms, receiving oral sex from their bar maids. Apparently this type of sexual encounter was the main attraction of the bar girls' entertainment package.

It was all too commercial, public, and rough for me. My fighting bar maid became irritated at me for not responding to her advances with an erection. She redoubled her efforts, which I resisted. We had a wrestling match until she had enough. She then started calling me a "sissy boy," their term for a homosexual. The real reason was that her aggressive approach combined with the entire bar's sex scene was the most unerotic experience I had ever had in my 23 years of life. Finally the brute left my booth in search of a more lucrative victim.

I could not leave that back room fast enough. Since there were still a couple of hours before the officers returned with the jeep, I decided to get drunk. I sat alone at the bar and got hammered, to the point that I almost passed out. However, for security reasons, I forced myself to stay conscious. Somehow I walked out of the bar and lay down on a bench near the club's door to wait for the officers. When they finally did arrive, they were irritated because I was still too drunk to drive. So one of the officers had to drive me and his buddies back to the firebase. They gave me one of their M-16s; I was supposed to be the shotgun guard for the jeep. There was no way I could have been effective fighting off a VC attack. Instead I just lay in the back of the jeep trying not to pass out. Happily, we made it to our firebase safely. That was the first and last time I was asked to drive officers to Saigon, which was fine with me.

Another trip to Saigon occurred when I accompanied a couple of firebase buddies who had to take care of some administrative business. Their business was completed by early afternoon, and then one of the guys suggested that we visit a friend of his who was stationed with the Brown Water Navy in the outskirts of Saigon. We all agreed and drove to a wealthy neighborhood of walled villas along wide tree-lined boulevards. Our destination turned out to be an attractive villa surrounded by eight-foot walls on the Saigon River. The style of the villa was more Mediterranean than Vietnamese. The massive building and its walls were light brown, both topped with a red tile roof. The villa's grounds included a medium-sized dock in the back on the river. Two small Navy riverboats were tied up at the dock.

There were a dozen sailors station at this idyllic river outpost. These men were living the good life. First of all, this luxurious villa did not look military, but rather it appeared to be the home of a wealthy man. They had hired a number of young Vietnamese women who cooked, cleaned and served the food. Most of the sailors had Vietnamese girlfriends, who visited them often. The largest room was set up as a bar-dining room, where the women served us a delicious American-style meal complete with beers at no cost to us.

Struck by the beauty of the surroundings and the non-military hedonistic feel of the outpost, I became jealous of these sailors. This type of civilian lifestyle was not unique for American soldiers and sailors fortunate enough to be stationed in Saigon. For example, many American MPs were billeted in air-conditioned civilian hotels in central Saigon. But as the day wore on, while we drank, smoked pot and socialized, I became more anxious. Even in this unbelievably idyllic setting, we were still in the middle of a war zone. Yes, they had a wall around the compound, but there was no guard tower or guardhouse near the front gate; nor were there any heavy weapons for defense. I felt vulnerable. As dusk approached I suggested that we leave soon. The American military dominated the air and the day, but the communists owned the ground and the night. I would not feel secure staying the night in that villa, and I definitely did not want to drive back to our firebase in darkness. My friends were reluctant to leave the luxury, but they finally accepted the wisdom of my suggestion. Just as the sun was setting we drove back to our firebase without any problems.

As I thought over our day, I was amazed over how peaceful it seemed to be stationed in Saigon. It always appeared as if the war did not touch the city. Of course, this was just an illusion; for example, I knew of the battle in Saigon during the enemy's 1968 Tet Offensive. But it was an attractive fiction to me. Saigon was an oasis for soldiers like myself who lived and fought in the far too ugly real war in the Mekong Delta.

25

Tan An Sniper

An incident that brought me close to death occurred in the South Vietnamese town of Tan An, which was seven miles south of our firebase near the village of Tan Tru. Our firebase was so small that it did not rate a PX, which is military speak for a retail store for American soldiers. To obtain various personal items like toothpaste or shaving cream, soldiers at our firebase had to travel the dangerous dirt road from the base to the junction with Highway 4 and then go south about five miles on the highway to Tan An.

Since I worked mainly at night, making the PX trip was always a problem. Skipping sleep one day, I wanted to start out early after finishing my all night duty shift at the TOC. As a safety precaution, our commander had ordered that everyone leaving the firebase had to be armed with his M-16 automatic rifle and wearing his flak jacket. It was always so hot and humid in the Mekong Delta that he did not require that we wear heavy steel helmets. Instead, I always wore a standard wide-brimmed bush hat as protection from the sun.

That eventful morning I ate a quick breakfast, loaded my M-16, put on my flak jacket and hung a bandolier of a dozen full M-16 magazines across my chest. As I was leaving, my sniper roommate asked me to buy him a carton of Salem cigarettes. He gave me money to cover the cost of the cigarettes and I double checked that I had my money. At the front gate I caught a ride on a deuce-and-one-half supply truck that was part of a convoy to Saigon. The description was not the weight of the truck. Army trucks were classified by the size of their maximum payload. Thus a deuce-and-one-half-ton truck could carry at most two and one-half ton of supplies. As I rode in the back of the truck, I kept my M-16 locked and loaded while I scanned the countryside for signs of the enemy. VC snipers routinely shot at convoys on this dirt road, and occasionally a VC squad would attack. Thankfully, on this day we were not attacked.

25. Tan An Sniper

At Highway 4 I dismounted, because the convoy was headed North and Tan An was located South. I crossed over the highway, stuck out my thumb, and in a few minutes I was picked up by two GIs driving a small one-half ton truck that would go by Tan An on its way to the 9th Division's base camp at Dong Tam. On this paved highway I was more relaxed because Military Police (MP) armored cars continually patrolled along it, while gunship helicopters patrolled overhead. Twenty minutes later the driver dropped me off near Tan An, which was a bustling town. It had a large roundabout in its center that enclosed a park. The sidewalks surrounding this central park were filled with Vietnamese civilians and American soldiers shopping at the various retail shops and motorcycle garages that lined the roads. I walked the one-half mile to the 2nd of the 4th Artillery Brigade's headquarters base at the edge of the town.

At the base the guard houses that flanked the front gate were manned with armed guards, but the gate itself was wide open. No one checked me as I entered. I simply walked through the gate and directly to the small PX. I wanted to return to my firebase as quickly as possible so that I could have a few hours of sleep before reporting for duty at the TOC that night.

At the PX I bought my personal hygiene products, a few candy bars, and the carton of Salem cigarettes. The clerk packed the goods in a standard American grocery store paper bag. The end of the carton of Salems stuck out from the bag. Carrying the bag in one hand and my M-16 in the other, I walked to the base's front gate and out into the town. The sidewalks were crowded with Vietnamese civilians and a few American soldiers.

At one alley a Vietnamese man approached me and asked if I would sell him American dollars. There was an active black market in all of South Vietnam for American currency—greenbacks. The currency market worked like this: an American soldier sold a $20 greenback to a Vietnamese for 40 or 50 dollars worth of Military Payment Certificates (MPC). American military personnel in South Vietnam were paid, not in greenback dollars, but in paper Military Certificates that looked like some type of Monopoly funny money. In an attempt to keep black market currency trading at a minimum, every few months there would be a surprise changing of the MPC funny money. These switches of the printed MPC would make the old MPC worthless. Only American soldiers were allowed to turn their old MPC in for the new ones. The American traitor soldier could buy money orders at an Army post office on some base with his ill-gotten MPC profits. Then he would send the money order home, asking his home contact to send him another $20 greenback bill in the mail. The traitor could then perform the illegal currency transaction again and again.

I never became involved in the black market for American dollars because I knew that our communist enemies were using those greenbacks to buy weapons and ammunition on the international weapons market. Those black market greenbacks were being used to kill American soldiers. So when that money trafficker approached me from his alley, I angrily cussed him out and walked away. Soon I was walking on the sidewalk that surrounded Tan An's central park circle.

It seemed like a normal afternoon, when BAM! A shot rang out. I knew by the bullet's whizzing sound that it was close. I dropped to one knee and turned to discover that it had hit an American soldier, who had been walking about three feet behind me, in the chest. I switched off the safety on my M-16 and brought it up to my shoulder as I searched for enemy targets. I was close enough to the victim that I was able to judge the bullet's trajectory; the shot had been fired from the rooftop of a two-story building located across the park and its surrounding roundabout. I desperately scanned that rooftop for the enemy sniper, all the time fearing he would shoot again. Deep down I wanted to retaliate, to spray that offending rooftop with automatic M-16 fire, but my better judgment prevailed. I was about 150 yards away with dozens of civilians moving around between the building and me. Spraying rounds at the rooftop in what we called "spray and pray" fire could possibly kill some of those civilians. Plus I never saw the sniper or any other viable enemy target. I did not shoot.

By this time five or six American soldiers had run over to help the sniper's victim. I heard one of the soldiers calling for an ambulance. I did not know the victim. His face was a pasty grey, full of pain and fear, but he was alive. He was just another young American soldier, walking unarmed and wearing the army's standard green jungle fatigues. At that instant I understood why I had not been shot, even though I had walked over that spot seconds before. Although it was not a genuine bulletproof vest, my flak jacket did provide some protection from distant rifle fire. I was a hardened target, more likely to survive a sniper's chest shot. Evidently the VC sniper skipped over me to wait for a softer target, the next soldier without a flak jacket. I stayed on my knee for another minute, knowing that by this time the VC sniper was running away from his hide. Then I stood up and walked away from the wounded soldier, who was receiving plenty of care.

As I walked to the highway I was still frightened, tense and jittery. Over and over I imagined the sniper's sights scanning over me as he searched for the right target. My mental images were so vivid that I felt the sniper's sights crossing over my chest. I kept my M-16 up in the ready position as I continued to search the rooftops for other VC snipers. I

25. Tan An Sniper

obsessively considered how close I had been to being shot or killed, which kept me upset and angry. I was so anxious that I have no memory of hitchhiking on Highway 4 to the intersection with the dirt road.

That intersection was one of our firebase's checkpoints. American soldiers were required to stay at the checkpoint until a large convoy of military vehicles would assemble. It was too dangerous for a single vehicle to drive along the dirt road to our firebase or for a single soldier to walk along it. Because the intersection was an assembly point, Vietnamese entrepreneurs had set up a small retail stand for drinks and snacks under a tree. A young Vietnamese woman ran the stand. Like all Vietnamese she was short, not over four foot six inches tall, and thin. She wore the popular white pajama-like outfit. I did not know her name, but she was infamous at our firebase for being a big time drug pusher. Along with her food and drinks, she sold opium and amphetamines. She was an aggressive dealer.

As I stood near the dirt road she spotted the carton of Salem cigarettes sticking out from my paper bag. In poor English she demanded that I sell her the carton of cigarettes. If they had been mine, I might have sold them; but because they were for my roommate, I refused. She persisted in trying to buy the valuable Salem brand of cigarettes that all the Vietnamese loved. She knew that she could resell them for a big profit. I continued to refuse and repeated for her to "di di" (go away) in my primitive Vietnamese. She only became more upset and demanding. Suddenly her English improved, as she proceeded to scream a barrage of obscenities at me. Her cussing was vile and abundant, depraved enough to make any tattooed sailor proud.

There I was, an American soldier in a foreign land, standing over six feet tall, with a machine gun in my hand, being verbally attacked by this little witch, shortly after almost being shot by a VC sniper. At first I was taken aback by the ferocity of her obscene barrage, but as I silently endured the onslaught, I became more and more irritated. Her obscenities continued as I struggled with my psychic vertigo from almost being killed to now having to listen to this little witch. The savage in me wanted to chop her in half with my machine gun. It was not just the sniper; all my anger and pain at being exposed to all the crazy evil of my entire Vietnam War coalesced against this she-devil. I thought, "All I have to do is lift my M-16 and with a slight pull on the trigger I can blow her away." The muscles of my forearm and hand tensed; my finger tightened against my rifle's trigger. It was a mighty internal struggle. The civilized part of me fought back heroically. I almost had to take my left hand to push down my right hand holding the weapon. Years of disciplined upbringing surfaced to force me to inhibit my baser drives. I just stood there silently.

However, there was a price to be paid for my self-control. For years my memory resurrected the incident, awakening the accompanying emotional pain and psychic conflict. After years of re-examination, I finally gained the insight that the power of my M-16 machine gun made all the difference in the psychological calculations involved in my decision not to pull the trigger. If I had been unarmed, I probably would have cussed her out or pushed her away. But possession of that machine gun in my hands escalated my violence potential. It was not about yelling back at her or pushing her. I was the God-King, the giver of life or death. That M-16 machine gun elevated me to a higher level of ferocity, that of lethal violence. It was not about pushback; it was about cutting her to pieces with an assault of machine gun bullets.

The trance-like emotional state of my internal struggle was shattered by a shout of, "McDermott, McDermott, get on the truck!" Distracted by the call, I turned and saw the artillery sergeant major, the one that all his men hated, calling me from the passenger seat of a deuce-and-one-half truck in a supply convoy. He may have recognized from my body language that my encounter with the little witch was volatile and could end badly. Or perhaps he was asserting the authority that he so coveted. Anyway, his shouts broke the tension. There was no way I was going to shoot the woman in front of witnesses. I made my decision; I ran to the truck and climbed up the back, still trembling with anger. We made it back to the firebase without incident. That night I kept my own counsel; I gave my roommate his cigarettes without telling him about my experiences. Later I smoked joint after joint of marijuana, trying to chemically escape the reality of the Vietnam War that was now my world.

Today, decades later, as I contemplate the stressful encounter, I know that I made the correct decision. My life would have been vastly different if my dark side had prevailed and I had killed her. Even if I had avoided prosecution, I might have been plagued by guilt; she could have haunted my nightmares for years. Yet I wish I had punched her; the witch deserved it. She was an incarnation of evil, another cancerous cell in the gigantic malignant tumor that was America's Vietnam War.

26

Religion

I was born and raised in a Polish village that happened to be located in the city of Pittsburgh, Pennsylvania. While I was growing up there in the 1950s, the city was enjoying a post World War II economic boom. My neighborhood was known as Polish Hill, a steep hill above the Allegheny River, roughly three miles from Pittsburgh's Golden Triangle. It was a Polish Catholic community that wanted to preserve the old world traditions; for example, our neighborhood blocked off the hill's main street, Brereton Avenue—where my extended family lived—so that church rituals like the Blessed Mother's May Day procession could parade up the avenue to the heart and soul of the community, the Immaculate Heart of Mary Catholic Church. This church was built early in the 20th century through the sacrifices of working class Polish immigrants, who wanted it to be a spectacular statement of their faith. Its design was intentionally modeled after St. Peter's Basilica in Rome. This edifice is still magnificent today, over a century later; it is an official Pittsburgh Historical Landmark.

My mother was of Polish descent, my father of Irish descent, and both were children of immigrants from two of the most Roman Catholic dominated countries in the world, Poland and Ireland. Both of my parents were true believers who lived their faith; for example, my mother insisted that all her children and my father, when he was home from work, kneel down together and pray the rosary along with a radio station that broadcasted the prayer every night at seven in the evening, except on Sundays. Of course, we all went to Mass every Sunday wearing our best outfits. I attended the parish's Polish grade school, where strict nuns, wearing their medieval-appearing habits with pride, taught us primarily the Catholic religion. Corporal punishment was swift and frequent.

I was totally brainwashed into the Catholic mindset. My stepsister and I would go to confession on just about every Saturday night so that we could receive Holy Communion the next Sunday at Mass. I was an altar boy from the third to the eighth grade, and also a choirboy from fourth through seventh grade. I remember one instance at school in which the nun in charge of the choir was arguing with the nun in charge of the altar boy schedule over whether I would participate in a High Mass at the altar or in the choir. I was a better altar boy than singer, so I was glad that the altar boy nun prevailed.

When I was a high school junior, my immediate family moved to the city's North Hills suburb of Shaler Township, where I attended the local public high school. Although we still attended Mass every Sunday, in this suburban community I was freed from the constant religious indoctrination, which allowed me to start thinking for myself. I read voraciously and began to develop religious doubts, especially about the eternal damnation of Hell. God's promise of eternal punishment in everlasting burning agony for committing just one mortal sin was standard Roman Catholic doctrine. Yet I wondered if a young man who simply ate a hamburger on Friday, which was a mortal sin at the time, really deserved agony for eternity. To me, it made God crueler and more vicious than a Hitler or a Stalin. What could a normal person, not a Hitler or a Stalin, do that truly deserved burning pain forever? Perhaps murder, I wondered.

Yet when I went off to the university I practiced the faith. I worked hard labor in a factory during the summers in order to pay my way through college. As a student who did not start studying in earnest until I entered my suburban high school, in college I wanted to prove to myself that I was intellectually capable. I kept an extremely strict study schedule, but I always found time to attend morning Mass on the day of a big test. Feeling that I had God on my side reduced my test anxiety. Yet I was not relying on miracles; I studied intensively hour after hour for each test. The combination worked, because I graduated in three and one-half years with highest honors.

During my first week in basic training, I attended the Sunday morning religious service open to trainees. It was not a Catholic Mass; instead it was a Protestant-type service with communal singing of hymns and Bible readings. I was disappointed; the service was not the religious experience I had expected. Later that week I learned that on Sunday mornings, I could use the few hours off to socialize over beers with my training mates at a base PX. This break from the military regimentation was the lifeline I desperately needed. I never attended another military religious ceremony in the States again.

The humiliation and harshness of Vietnam era military training discipline nearly broke my spirit. As I detailed in the chapter on basic training, I was so desperate that I applied for conscientious objector status. As part of the process, I was dismissively sent to an Army chaplain. He was clueless; he even asked me why I was afraid of him. To me, he was not a priest or a minister, but rather just another Army officer. I was given my conscientious objector paperwork and sent on my way to AIT.

While I was home on leave after basic training, my father, who knew that I was struggling with my conscience over the prospect of killing, took me to a real Catholic priest to discuss my problems. The priest was a young man who had no real insight into my dilemma; he kept on spouting religious platitudes that were totally irrelevant and useless. The meeting was a total waste of time. My depression grew more severe.

In my basic training letters to my sweetheart I had mentioned thoughts of suicide. Those feelings resurrected the night before I was to return to the Army for AIT. I was driving back from my sweetheart's home on a two-lane highway, where trailer trucks zoomed by in the opposite direction just feet away from my car. In my despair I thought how easy it would be to turn my car into the path of one of these huge trucks to end all of my misery and pain. The lure of the peace of death enticed me for a moment, but then in my negatively warped thinking I concluded, "With my luck I probably would not die, but instead end up horribly disabled or paralyzed for the rest of my life." I saw even more pain in my future; attempting suicide was not the easy answer.

As I gathered more information about the conscientious objector process, I discovered that one of the primary questions the Army psychiatrist was sure to ask was, "If someone were shooting at you to kill you, would you defend yourself by shooting back?" For me this question was an extremely serious matter of conscience. I was never one to delude myself with falsehoods; I had to be totally honest with myself. I seriously contemplated and reconsidered the question over and over. What would I do in that situation? I recalled how as a boy I had often been attacked on the mean streets of Pittsburgh. Not always, but more than once, I had fought back.

After much internal struggle I came to the honest conclusion that self-preservation won out. Rather than following some ancient religion's rules, I wanted to live. In a kill-or-be-killed scenario, I would fight back to save myself. It was a monumental insight that changed the course of my entire life. Of course, I did not have the luxury of answering to the psychiatrist that I would do my best to avoid such potentially lethal situations. That option was no longer a viable one for me because I was already in the Army and on my way to Vietnam one way or another.

To be honest, I predicted that, if I were granted conscientious objector status, in all likelihood I would be sent to train as a medic before being ordered to Vietnam. I knew that serving as a medic in the Vietnam War was an extremely hazardous job. This atrocious war was one in which our communist enemies would deliberately shoot to wound a soldier and then wait for the medic to come to the wounded man's aid. Then the communists would shoot to kill the medic. They understood fully how the loss of such a vital asset would diminish the combat effectiveness of any American platoon. On my trip to Fort Sill for my advanced training, I took all my conscientious objector paperwork and threw it into an airport trash bin. I never brought up the issue again, and military officials never mentioned it afterwards.

Once I reached South Vietnam I was required to spend that first week at the 9th Division's primary base camp in Dong Tam, near the town of My Tho. There I was issued and trained on my fully automatic M-16 assault rifle. The training was also designed to acclimate me to the high humidity and heat of the Mekong Delta. My basic fear and anxiety grew worse now that I was in the war zone and heard the frightening stories told by the veterans.

That first Sunday I had time off to attend an actual Catholic Mass. Before entering the room used for the service, everyone was required to check their weapons into a weapons room, which was like a coatroom for guns. While checking my M-16 in, I studied the gunroom. It was weirdly fascinating. It was a gun aficionado's dream room, literally overflowing with the widest assortment of individual weapons that I had ever seen, before or since, in one place. Weapons hung on all four walls from the floor to the ceiling. They were also stacked in rows of gun racks that crowded the floor. Of course, there were many standard-issue American military weapons like the .45 caliber automatic pistol and M-16 rifles, but there seemed to be more exotic weapons than standard issue ones. There were American military weapons from previous wars like M-1 carbines, Thompson submachine guns and Browning Automatic Rifles (BAR). There were dozens of different types of foreign weapons, like the Canadian-made Browning 9-mm automatic pistols and Russian Makarov 9-mm pistols that were possibly captured from the communist enemy. As could be expected, there were dozens of AK-47 assault rifles and many intriguing foreign weapons that I did not recognize. They all appeared so sinister, gathered together radiating evil in a room next to a church room. The gunroom was incongruent with my expectations of the aura of a religious service. It was too bizarre for me.

26. Religion

The religious service was a Catholic Mass, but it too was strange. For one, the chaplain/priest wore the standard olive-green jungle fatigues we all wore, rather then the sacred vestments worn by civilian priests. Even though I do not recall the details of the chaplain's sermon, I do recall my impression that he was preaching as if this war was acceptable. My mind at the time was reminded of a popular slogan expressed among cynical Vietnam combat veterans: "Kill a Commie for Christ." It was much too outlandish and peculiar for my sensibilities; I never attended another military religious service, Catholic or otherwise, while in the Army.

It did not help my faith that my primary assignment was to the artillery firebase near Tan Tru village. We did not have a chaplain during most of my tour there. There was a chaplain when I first arrived, but I remember him hosting sporting events like basketball games and boxing matches rather than religious services. That chaplain was rotated back to the States and, for whatever reason, never replaced. No one held any religious services at our firebase. The other limitation was that no one received any days off at our firebase. The only exceptions were the highly prized individual weeklong rest and recreational (R&R) leaves. There was no time off on Sundays for religious services. We were always at war; we all had duty every single night and day. The VC or the NVA could attack our firebase at any time. Everything else in our lives was secondary to the war.

Towards the end of 1969, some military authority ordered a Christian chapel built near our headquarters bunker. It was a small one-room building that could not hold more than 30 people. It was one of the few buildings at our firebase that was painted; it was white and appeared quite delicate. The structure puzzled me because it was never utilized for any services. It was always open. I think a few men may have gone there to pray occasionally or to escape the military regimentation. Its purpose and its small size is still a mystery to me.

Over my 12-month combat tour the gruesome realities of the Vietnam War affected my attitudes and thinking about religion. Yet I had had religious doubts even before Vietnam. While in basic training I had written to my fiancée (July 11, 1968), "I pray all the time, trying to answer questions like why a just God can allow this war to happen." I heard of atrocities like the Viet Cong communists painfully torturing a captured American enlisted man and then mutilating his dead body. Just like the millions of Polish Jews in Hitler's Nazi extermination camps, I asked myself, "Where is God?" Although I never personally witnessed these war crimes, there were multiple reports of patrols finding missing American soldiers tied to trees, dead from bayonet wounds with their penises cut off and shoved into

Small chapel built near our living quarters bunker.

their mouths. Where was God when the VC peeled the skin off of live captured American soldiers? The biased world media never reported on these communist atrocities, only on American misdeeds. How could a just God allow the Viet Cong to attack our hospital in Cam Ranh Bay, where they killed women nurses as well as the wounded soldiers? At least the *Stars and Stripes* newspaper reported on that atrocity. To me it was a more than a vexation that God, who I was taught was omnipresent, did not seem to be in Vietnam. The principles of the Catholic catechism as taught by the nuns and priests back in Pittsburgh did not encompass these grotesque horrors of the Vietnam War.

There were no communist prisoner of war camps in South Vietnam.

The Viet Cong tortured and killed all their prisoners of war in the South. There were only a few prison camps in North Vietnam. The only American prisoners the Vietnamese communists kept alive were officers who were pilots or navigators shot down and captured in North Vietnam. There were also just a few enlisted men, who were tail gunners on our B-52 bombers. It is well documented that these unfortunate American officers were horribly tortured for secret war information. Every one of the officers was broken, which resulted in these men suffering crippling guilt after their repatriation. How could God allow this extreme torture happen? Where was God when the VC raped and killed the daughters of Vietnamese fathers who were resisting the communist's efforts to draft their sons into their army? The VC raped and killed these young women in front of their fathers and then killed the father as a message to other South Vietnamese fathers who had military age sons.

Often I have heard veterans of America's previous wars firmly declare that, "There are no atheists in foxholes." That statement completely disregards the fact that our Vietnamese communist enemies were atheists. The closest thing to a God for them was their supreme leader Ho Chi Minh, whom they venerated even after his death. My argument is that a horrific war like the Vietnam War can change a man into an atheist. Anyway, if I am in a foxhole with a man of faith, I want him to be shooting at the enemy, not praying. I worked at the Jacksonville Veteran's Administration Clinic with a true World War II hero, Dr. Harold Baumgarten, who fought and was seriously wounded in the bloody battle of Omaha Beach on D-Day. He told me that during the invasion he saw one terrified American soldier kneel down and start praying with his Catholic rosary beads on that terrible beach. Immediately German machine guns cut the supplicant in half. That particularly symbolic death is depicted in the vivid opening beach combat scenes of Steven Spielberg's film *Saving Private Ryan* and is based upon recorded interviews with Dr. Baumgarten.

27

Weddings

The headquarters of General Westmoreland and his staff was located in South Vietnam's capital city of Saigon. The supreme command's official name was Military Assistance Command Vietnam; however, among American troops it was mainly referred to by its acronym, MACV.

MACV did not encourage fraternization between American soldiers and Vietnamese civilians. What were especially frowned upon were marriages between enlisted men and Vietnamese women. Evidently high command could not prevent such marriages; instead, they would generate administrative obstacle after administrative obstacle to discourage the enlisted man and his Vietnamese woman.

Over the first few months of my combat tour at the Tan Tru firebase, I made the acquaintance of an artillery clerk who was attempting to defy MACV and marry a Vietnamese woman. He never disclosed how he had met the English-speaking woman and fallen in love; however, he did repeatedly complain about the many administrative obstacles he had to overcome to marry her. High command had demanded so many requirements to meet and forms to complete that he had been forced to volunteer for another one-year Vietnam tour of duty to finish the task. In my opinion it must have been real love for him to risk his life with a second year of Vietnam War duty in order to marry his Vietnamese woman.

After I had socialized with him for about six weeks, this fellow happily informed me that he had successfully met all the regulations and completed all the required paperwork. Soon after, they were married somewhere off base, in the women's hometown I suspect. But even then, as an enlisted man, he was not allowed to live with his new wife; he continued to live with his bunker mates on base. Despite that rule he was able to somehow get her a job as a barmaid at our firebase's NCO club. This arrangement enabled them to see each other daily and on occasion enjoy conjugal visits.

27. Weddings

Not aware of arranged special signals, I accidentally walked into his bunker room once and interrupted one of these special visits. Instantly perceiving my faux pas, I quickly withdrew, apologizing profusely.

About two months later, while I was on duty, an artillery sergeant informed me that the clerk's new wife was dead. He explained that recently she had traveled alone to Saigon. While in the capital she had been arrested by the "white mice," which was our epithet for the South Vietnamese police because of their small stature, white uniforms and vain attitude. The police had accused her of spying for the communists. Incredibly, the next day she was tried and found guilty. The following morning she was executed.

Stunned by this turn of events, as soon as I was off duty I hunted down the widower. As I feared, I found him in his bunk utterly crushed by his grief. He was in the process of drinking himself into unconsciousness. I offered words of condolence, but at that point he was beyond consolation. He simply continued drinking.

To this day I do not know if his wife had been a spy. I knew the widower well enough to know that if he had any hint of suspicion, he would never have married her. Yet I cannot help but think: What better job for a female spy than a barmaid at a forward firebase's NCO club? She was in the perfect position to gather valuable intelligence for the enemy. She would have been able to overhear tipsy non-commissioned officers discussing vital topics like future operations. It may have been that the VC, upon learning of her new job, turned her into a spy by threatening to kill her, along with her entire extended family, if she did not cooperate. The Viet Cong were well known for committing such atrocities.

As one would expect, his wife's sudden execution devastated my friend. At times I feared that he would kill himself. One of the many aspects of the story that irritated me was how the Army had forced him to extend for a second tour; now he was forced to endure eight more months of Vietnam War duty. He was thus deprived of the comfort of returning home to cope with his grief in the supportive arms of his family. Instead he chose to numb his grief through the slow suicide of heavy drinking. He never again was the man I used to enjoy spending time with.

MACV high command not only discouraged marriages between enlisted men and Vietnamese women, they seemed to discourage all weddings of Army enlisted men. As I heard one sergeant explain: "If the Army wanted enlisted men to have wives, it would have issued them one." When I tried to make arrangements for my wedding while only an E-4 Specialist, I ran into the Army's impediments against such marriages.

On my leave before being sent to Vietnam, I proposed to my college sweetheart. She accepted and wore my diamond engagement ring as we kissed good-bye at the airport. We wrote to each other daily. Through our letters we decided that she would come to Hawaii, where I could meet her on R&R, and there we would marry. We also decided that the best time for us was August; consequently, I applied in the spring for an August R&R leave to Hawaii. My application for August was rejected; any leave would have to occur in September.

So I wrote to my fiancée to make her arrangements. I also wrote to the Army's lawyers, the Judge Advocate Generals (JAG) in Saigon, inquiring about the rules and regulations for getting married in the State of Hawaii. They in turn wrote to the chaplains in Hawaii for the regulations. JAG put me in touch with Hawaii-based chaplains, who explained that I could not meet the legal requirements out of state. We would have to get the blood tests and marriage license in Hawaii.

My R&R leave application for September was rejected for a reason that was never explained. At this point I was becoming desperate, especially after having to inform my fiancée that she would have to change the date of her airline tickets and her other arrangements. She was upset. The cost of the trip and the changes in schedule were expensive. Money was an issue. I needed to have some help in dealing with the Army, so I asked my West Point graduate captain, who was the battalion's FO, if he would aid me. He agreed and we both traveled to division headquarters at Dong Tam, where his presence and support worked. I was guaranteed an October date for my R&R. As soon as I could, I wrote to my fiancée to give her the good news and tell her to make her reservations for October.

Weeks later I was finally aboard an American commercial jet liner flying to Hawaii along with over 100 married soldiers, everyone looking forward to meeting his lady. Hawaiian R&Rs were usually reserved for married soldiers so that they could meet their wives on the resort islands. Our jet plane landed in Honolulu and we taxied up to a large group of waiting women. The plane stopped and we all disembarked as fast as possible. All the guys were running to their women. The couples paired up as I looked for my fiancée. I searched and searched through the faces of the women. My fiancée Judy was not there. No one was waiting for me.

It was an awful blow to my psyche. I was psychologically reeling as I walked toward the city center. After thinking over my situation, I convinced myself that I was not going to allow my fiancée not appearing to ruin my R&R. So I walked into the nearest bar and ordered a cocktail, drank it, and walked to the next bar. I walked in and ordered a cocktail, drank it,

27. Weddings

and walked to the next bar. And so on and on; by ten in the morning, I was a happy drunk. Strolling along the sidewalk, I spotted a motorcycle rental lot across the street. Up to that point in my life, I had ridden a small motorcycle only once during college. I was not exactly Easy Rider. But in my addled state, the idea of driving a motorcycle was appealing. I rented a powerful street motorcycle.

I successfully drove the beast off of the rental lot out into the street. I drove down the street for a block before stopping for a red light at the first intersection. I was the first vehicle at a standstill in front of the red light. To balance the motorcycle, I had to stand with my both of my feet on the street. When the light changed to green, I turned up the throttle and popped the clutch. The beast shot out from under me, rolled on for a few yards, then fell over, bouncing on its side while its wheels spun.

I ran up to the motorcycle. Its handlebars were damaged, but I was able to pick it up to push it out of the intersection. Having traveled only a block, I could see the rental lot; I pushed the damaged motorcycle back to it. The shop owner came running out yelling that I would have to pay, because I had signed the contract. "What contract?" I asked. It was the rental agent's first day on the job and he had neglected to have me sign the rental contract. I told the owner not to fire the guy because he had helped a soldier on leave from fighting in Vietnam.

Having dodged a big expense, I decided to walk off the alcohol's effects. So I walked to the USO welcome center near Waikiki Beach. Once there, I called my cousin JC, who was to be my best man. My cousin was a Coast Guard medic who was assigned to the Navy hospital in Honolulu. This handsome bachelor lived an idyllic existence, especially considering that he was in the military. He picked me up in his sports car with a surfboard in the back seat. Before leaving the center, I left a note on the big message board set up in the building on the slim hope that Judy would read it. We drove to the apartment he shared with two other Coast Guard sailors, which was located only a few blocks from Waikiki Beach. After hearing my sad story, JC wanted to cheer me up. He suggested we walk down to the beach. I agreed that was a great idea.

Waikiki Beach deserves its favorable reputation. With a beautiful view of Diamond Head's volcanic cone, the tropical beach was crowded with pairs of young women enjoying the luxury. Amazed by the beauty and the peace of the beach, my mind shot back to Vietnam. How could all these people enjoy themselves here? Didn't they know that there was a war going on? Did they not realize that American boys were dying on the other side of the ocean? For the first time I truly understood that for most Americans the Vietnam War was an unpleasantness that did not affect their lives. It was a harsh reality that was difficult for me to accept.

Meanwhile, JC was walking down the beach smiling and greeting the sunbathing women. If the target of his greeting did not immediately smile back and respond, he simply walked on to the next pair of attractive women. Within 20 minutes he had made a connection with an attractive pair of secretaries, who agreed to meet us that evening for drinks and dancing. For the first time on my long anticipated R&R leave, I was pleased.

We returned to his apartment. While hanging out there, I received a phone call from my fiancée. She had read my note at the welcome center. She was staying at the apartment of a sorority sister who was teaching on the island and who was to be her bridesmaid. The Army had told her that I would be flying in the next day. These many years later I cannot but wonder if that wrong date was a deliberate fabrication or simply another Army administrative blunder. We had a joyful reunion at my cousin's apartment and then moved to the Sheraton Hotel on Waikiki Beach. It is a grand old hotel; we stayed in a brand new tower built onto the main building in an upscale room. The beach in front of the hotel was postcard perfect with its grand view of Diamond Head and the beach. Here the waves broke 300 yards offshore, perfect conditions for surfboarding.

We spent the next two days completing our application for a Hawaiian marriage license and getting our blood tests. My cousin JC agreed to be my best man and Judy's sorority sister, Mary Jo, agreed to be the bridesmaid. Judy arranged for flowers and a professional photographer. We were required to meet with an Army chaplain, who asked, "What religion are you?" Judy answered, " Catholic," while I answered, "Protestant." We looked at each other. Without consultation we had both switched to report the partner's religion. Quickly I took command and insisted: "Protestant." The Catholic chaplain responded, "Thank the Lord for Martin Luther. I am all booked up today."

The Catholic chaplain sent us to an Episcopalian chaplain. He recognized at once that we were mature young adults. We were both college graduates who had known each other for over four years. He agreed to marry us on our third day of R&R, but first he gave us a brief Pre-Cana counseling that he claimed he had cut down from four hours. The chaplain counseled: "Always remember and apply the Five Cs of a Happy Marriage: (1) Cooperation, (2) Communication, (3) Consideration, (4) Compromise, and (5) Conjugal Bliss."

The chaplain then scheduled our wedding in his last slot of the next day. I appreciated this scheduled time because it meant that we would not

27. Weddings

be part of assembly-line weddings. The chapel was performing multiple weddings every day. R&R soldiers wearing Bermuda shorts and sandals were marrying women they had just met. Being last in line meant no other couple would push us to hurry.

October 16, 1969, Day Three of R&R. In the morning I went surfing on Waikiki Beach; JC had taught me the basics. The surfing was excellent that day, with six-foot waves breaking hundreds of yards offshore. In the early afternoon Judy pulled me off of the surfboard to dress for the wedding ceremony. Judy's mother, a professional seamstress, had designed and stitched a beautiful white wedding gown with a knee-length skirt and a matching veil. Judy had brought my one and only brown college graduation suit for me to wear.

An Episcopal priest married us in Fort DeRussy's chapel near Waikiki Beach. Judy was breathtakingly lovely as she glowed walking down the aisle in her white wedding gown, carrying her orchids. She joined JC, Mary Jo and me at the altar. The sanctity of the chapel added to the beauty and solemnity of the ceremony. We made our vows and exchanged white gold rings. The photographer snapped remarkable pictures during and after the ceremony. After all the waiting, all the anticipation, we were finally husband and wife. Later our bridesmaid and best man took us to a fancy restaurant for a celebratory wedding feast. They toasted our future with champagne. The entire wedding day was a memorable success.

The next two days of my R&R flew by as Judy and I enjoyed our introduction to the Hawaiian culture. We discovered the luau, macadamia nuts, and Hawaiian shirts. We attended a parade and anniversary ceremony for Queen Ka'ahumanu, the woman leader who brought Hawaii into the modern world. My short recreational leave ended much too soon.

One of the most difficult things I ever did in my life was to leave my new bride in Hawaii to board the jet airliner that would fly me back to the Vietnam War. I hated parting from her and I hated returning to the war. We kissed passionately and promised to write. I could not look back as I reluctantly boarded the plane. Our one consolation was that I had only three more months in Vietnam before my tour was over. Once back at the Tan Tru firebase, I was a "short timer" who could start looking forward to being discharged from the Army.

28

Firebase Drinking Clubs

Since alcohol was by far the dominant recreational drug of choice for American soldiers in Vietnam, its story at our firebase is a necessary one to fully understand my war. It was the only drug officially sanctioned and sold by the Army. Every American Army base of size in South Vietnam had its officers', non-commissioned officers' (NCO), and enlisted men's club that sold inexpensive liquor and beer. Each of these three clubs at our firebase has its own interesting story.

One enemy attack clearly demonstrates how critical alcohol supplies were to the normal functioning of our firebase. On September 9 the VC blew up a small bridge by floating an explosive charge down the river, which hit the bridge's support beams. That bridge had linked the dirt road that led from our firebase to Highway 4, the only paved highway in our area. South on Highway 4 was the direct route to our division's headquarters in Dong Tam; North on the highway led to Saigon. The bridge's destruction effectively isolated our firebase. Wrecking it prevented supply trucks from Dong Tam or Saigon from reaching us. Consequently, vital artillery rounds could not be delivered in the normal manner, by truck. The subsequent ammunition shortage was so desperate that our commanders diverted limited helicopter blade-time from direct combat roles to resupplying the essential artillery rounds. This was only a temporary solution because the tonnage of artillery rounds delivered expensively by helicopter would never equal the normal tonnage delivered cheaply by Army supply trucks.

The blown bridge also prevented the resupply of beer and liquors to our firebase's three drinking clubs. Soon our firebase's alcohol supply ran dry. This deficiency was the cause of much concern among the troops; the alcoholics were beginning to suffer symptoms of withdrawal. Their desperation motivated senior NCOs to think creatively. One of them devised a unique solution to the alcohol crisis.

28. Firebase Drinking Clubs

One two-and-one-half-ton supply truck was loaded with beer and liquor at the Dong Tam warehouse. It was driven to one side of the blown bridge, which was now nothing but bent metal rubble across the waterway. In a well coordinated maneuver, two deuce-and-one-half-ton trucks drove from our firebase to the other side of the bridge's rubble. One of these trucks was empty; the other was carrying a 12-man work squad. Once at the bridge, the men of the work detail climbed over the bridge's rubble to form a human chain. Then the cases of beer and liquor were removed, one at a time by hand, from the Dong Tam supply truck to their end of the human chain. The man receiving the first case turned and handed it to the next man in the chain and so on until the case was safely deposited into our base's supply truck on the other side of the water.

The hard work continued until the entire supply of Dong Tam alcohol was transferred to our supply truck. The men of the work detail crawled off of the bridge and onto the other supply truck to return to our firebase. That night there were celebrations in all three of our firebase's drinking clubs. A solution had been found for the alcohol resupply problem until the bridge would be repaired, which happened within a week. Ammunition supplies were temporarily curtailed because of the destroyed bridge, but our Army did have its priorities. The beer and liquor got through.

Enlisted men's clubs tended to be large and unruly. On my first night in Dong Tam I met an old high school buddy, who took me to the base's crowded enlisted man's (EM) club. At the club I saw a soldier drinking a beer with a VC ear floating on top of the beer. It was a nauseating sight that brought home the reality that I was no longer in the world; I was in the Nam. The EM club at our firebase was equally wild. A beer or full shot of liquor cost only 25 cents. A private could go to the club with two dollars and get totally inebriated on the eight drinks his money would buy. Many soldiers regularly drank their fill. I rarely visited the EM club because I was never a big fan of alcohol or the craziness associated with intoxication. Mainly I went to the club for the entertainment; free movies were occasionally shown there.

One night the NCO who managed the EM club decided to show the popular movie *Green Berets,* starring the Army's favorite movie star, John Wayne. He was an icon for the American Army in the Vietnam War era because of his war movies like *Iwo Jima,* where he portrayed the quintessential war hero. Because of Wayne's roles as a courageous soldier or sailor, he was admired as the embodiment of the Army's fundamental values. His name became synonymous with gung-ho, aggressive behavior. Whenever a soldier did something outlandishly courageous, like charging a machine gun nest, the soldier was said to be "John Wayning it" by his peers.

I am not a John Wayne fan, mainly because he was such an incompetent actor. He really only portrayed one role, himself, no matter what character he was supposed to be playing. I also knew that his war exploits were just make believe. He was not a true combat hero like Audie Murphy. During World War II, Wayne was never in the military; many accused him of dodging the draft. When I heard that the EM club would be playing John Wayne's Vietnam War movie *Green Berets*, I refused to attend. However, the next day many of my buddies, who had attended, told me that it was so unrealistic that they regarded it as a laughable comedy. In fact, the movie proved to be so popular as a comedy that the EM club sergeant decided to hold it over for an encore performance.

So that second night I went to the EM club to watch the comedy. The club's NCO was projecting the movie outside the club in an open field covered with folding chairs for the audience. My buddies were correct to a point: the movie was an outrageous travesty, a comedy, but I did not laugh. It was simply wrong. For one thing, the foliage was wrong, since the movie was filmed in South Carolina and Georgia. John Wayne was fighting in pine forests, not the tropical jungles of South Vietnam. He was miscast; here was a 60-year-old obese man lumbering around, pretending to be a Special Forces colonel fighting in close-order combat. It was absurd casting, just like all the Japanese actors who played Vietnamese soldiers in the movie. The screenplay was poorly written, full of clichés and unrealistic dialogue. Here is Wayne's character talking about justice: "Out here, due process is a bullet." The movie's NVA officers were shown living in luxury and drinking champagne, which was far from the truth. Hollywood movies never portray war accurately; the reality of war is always more appalling.

I recall thinking to myself that it would serve us right if we were attacked while watching this terrible war movie. It was as if the VC were reading my mind, because about ten minutes later enemy mortars attacked us. The first rounds struck outside the berm, but after someone yelled, "Incoming," everyone in the audience scattered to safe cover.

One of my sniper bunker mates, who was scheduled to rotate back to the world in a couple of days, went to the movie wearing brand new fatigues, looking all clean and fresh. During the attack, he panicked and dove under the EM club building. Later my bunker mates teased him because he returned covered head to toes in foul smelling muck.

During the mortar attack I ran to hide behind a nearby supply truck, a hide that enabled me to observe the area. The enemy's mortars were striking

28. Firebase Drinking Clubs

Craters from enemy mortar shells exploding outside our firebase's berm during the rainy season.

nearer and nearer to our barbed wire barriers, but to my knowledge none of their rounds struck inside the berm. From my vantage point I saw that, for a few moments, John Wayne was play-acting the Vietnam War on the screen in the middle of the real Vietnam War. *But no one was watching.* All the folding chairs were empty. It was the most surreal scene of my Vietnam tour, a kind of poetic justice. What if they staged a war and nobody came?

In July the EM club was the site of another outrageously wild happening. This night the EM club was showing pornographic movies. The news spread like wildfire; every enlisted man who could do so planned on attending the show. I went to the club early in the evening, yet it was already tightly packed with soldiers. The pornographic movies were a disappointment. They were in black and white and were of such poor quality that they must have been filmed in the late 1940s. Their style was old fashioned; the women all wore nun habits and all the characters wore Lone Ranger–style facemasks. To me, the movies were not exciting; however, most of my fellow enlisted men were so erotically deprived that they enjoyed the show. As could be expected, there was heavy drinking, much loud hooting and jeers, and widespread horsing around. It was not my kind of scene; I left after watching for only a few minutes. On the way out I noticed that a growing crowd of men were standing outside the overcrowded club, watching the screen through the doors and cracks between the side boards of the building.

Our EM club was a small one-story structure built on piling supports and a wooden floor. The walls were only one-inch-thick planks nailed to the frame. Some of the enlisted men crowding the outside discovered that they could pull the ends of the siding away from the frame to create more viewing space. They were starting to do so when I left to return to my bunker. The next morning I was told by one of my bunker mates that as the night wore on the crowd became more and more unruly and destructive as they continued to drink heavily. Men started to pull the siding boards completely off of the building's frame to see the screen better. Others saw them pulling off the siding and decided that it was a great idea. Soon there was a frenzy of men pulling off siding with their bare hands. There was so much destruction that it weakened the club's frame structure to the extent that the entire building collapsed. I am not sure exactly how, but apparently the falling roof and framing did not seriously injure anyone.

The next day I returned to the site for my own inspection. It looked like a vicious wrecking ball had destroyed the building. All that remained of our EM club was wooden rubble and metal roofing material piled high on the floor. The colonel also inspected the site that day. I was told that he became angry as he viewed the wreckage. He was so upset that he vowed he would never rebuild the EM club. By pulling the EM club down, the alcoholic drunks had deprived all the other enlisted men of a favorite social outlet.

Afterward, when I desired the occasional alcoholic beverage, I would have to pretend to be a sergeant by borrowing an E-5 sergeant's fatigue shirt to wear in the NCO club. One night at the NCO club I was drinking alone at the bar when a sergeant that I did not know came in, saw me, and started yelling that I was not an NCO. He walked up near me and started pointing at me while screaming his message. His efforts to kick me out ended when the relaxed bartender told him, "Calm down, he is just sitting there having a drink." I appreciated the kindness, but I quickly finished my drink and left. That was the last time I went to the NCO club.

28. Firebase Drinking Clubs

My favorite NCO club story is that one night late a sergeant was walking near our bunkers on his way from the club. As the tipsy sergeant passed underneath, the squirrel monkey that lived in the roof of the MP bunker next to ours jumped down on his shoulders. The sergeant screamed out in terror, thinking that a VC had jumped him. His wails awoke sleeping men, who saw the reality and started laughing at the sergeant's predicament.

Finally there was the officers' club, standing alone in the middle of a large rice paddy near the TOC. I never went inside, but I was told that it was well furnished, with a high quality sound system and a large television set. It was popular with all the officers, including the colonel, who drank there regularly.

During the dry season, the VC captured a 106-mm recoilless rifle from an ARVN unit. A recoilless rifle is a lightweight, direct fire, artillery-sized weapon. It is recoilless because it has a recoil compensator that allows some of the gases from the fired propellant to escape out of the rear of the breech. This back blast is dangerous; any soldier behind the recoilless when it fired would be badly burnt. Its light weight made the weapon more portable than a comparable artillery piece of the same caliber. This weight advantage made it highly desirable to the VC.

Our firebase's officers' club was the primary target for a series of communist attacks with the newly captured recoilless rifle. The VC's first recoilless rifle rounds struck the rice paddy near the officers' club in the middle of the day. At first the senior officers assumed that it was a limited random attack. However, the next day a few more rounds were fired at the club; one struck a corner of the building. The enemy's firing of the recoilless rifle during the day made it difficult to spot the weapon's location. The VC knew that if they fired it at night, the light caused by its back blast would reveal its location. They seemed to have the officers' club accurately in their sights. For the next five days recoilless rounds were fired, some of which hit one corner of the club. The VC may have possessed intelligence on its exact location. Some officers blamed the enemy's accuracy on the fact that the club was directly in line with the tall radio tower that stood next to the TOC. There was a red light on top of the tower as a warning signal for helicopters flying to our firebase. These officers claimed that the illuminated tower acted as the ideal aiming post for the enemy.

These daily recoilless attacks became a serious irritant to the colonel and all the officers who visited the club. The VC were consistently striking one corner of the club where the television set had been sitting. Once, as the first mortar rounds hit, officers were seen diving out of the club's windows right into the muck of the surrounding rice paddy. After a week of the recoilless rifle attacks, the colonel finally became incensed enough to order a full-scale campaign to destroy the VC's recoilless rifle.

The worst problem for the colonel's destruction campaign was discovering the exact point where the VC were firing the rifle. The colonel devised a plan which required an observation helicopter circling our firebase continually during the daylight hours. On board, spotters using binoculars were to search for the weapon. For the next few days it appeared that the circling observation helicopter intimidated the VC, because they did not fire their weapon once, but eventually the VC's desire to kill officers overcame their caution.

Again the officer's club was the target for the enemy's recoilless rifle rounds. After a couple of days of the renewed recoilless fire, the observation helicopter succeeded in spotting the rifle's back blast. The pilot quickly called in Cobra gunships that attacked the VC position with 40-mm rockets. The rockets destroyed the recoilless rifle and killed the VC crew. There was a raucous and rowdy celebration in the officers' club that night. They could finally relax and enjoy their drinking and socializing in peace. They could make the necessary repairs to their club building without the worry that the structure might be destroyed the next day.

29

Viet Cong Attacks

The Viet Cong attacked our firebase repeatedly over my year of combat duty. At times the attacks would consist only of small arms fire. For instance, on the night of August 14 we were subjected to a surprise attack when the VC set up outside our berm and shot automatic rifles and machine guns, not at specific targets, but rather randomly at the entire firebase. I had been relaxing in my bunk when the attack started. Someone yelled out, "Incoming." A few of my bunker mates and I ran to one entrance to see what was happening. All that we could see were the enemy's green tracer rounds whizzing through the air. We could hear our perimeter guards returning fire from our firebase's perimeter guard bunkers.

It seemed to me to be a limited attack, but we could never be absolutely sure. The worry we all had was that the firing could be the opening salvo for a sustained ground attack. I was worried enough to take my M-16 down from the wall where I stored it. I had not fired the rifle or cleaned it in weeks. While loading the weapon, I started to worry that my rifle might not work. To ensure that it could still function properly, I took it to the other entrance to our bunker on the opposite side from where my mates were standing. I paused at the opening when I saw more of the enemy's green tracers bouncing off the wall of the next bunker. None of the rounds were penetrating the stacked ammo boxes, but there were dozens of tracers shooting through the air. It was intimidating. I chambered a round and fired a few shots into the dirt road near our bunker. I felt reassured when my M-16 worked perfectly.

However, my rounds panicked my bunker mates at the other end of our bunker. I had not warned them that I was going to shoot. I heard one of the guys yell out, "Gooks inside the wire!" which meant that the VC had breached our perimeter defenses and were shooting inside the base. If true, that would have been an extremely dangerous situation. I had to run to the other end of our bunker and calm them down by explaining what I had done. They were justifiably irritated with me and said as much; yet they were relieved that our perimeter was still intact. The attack lasted for a long 15 minutes as the VC shot hundreds of small arms rounds. Fortunately, there was no accompanying mortar fire, nor was there a follow-up ground assault on our base. Our battalion sustained no casualties that night.

Most of the VC attacks on our firebase used mortars as their primary weapon. The problem with a mortar is that it is difficult to be accurate with the weapon. In our Army we used forward observers who could see where the mortar rounds were striking and call adjustments to hit a specific target. The VC did not use forward observers; instead they simply lobbed the rounds at our firebase, hoping to hit something important. They were not always successful. On April 7 the communists fired 22 mortar rounds in the direction of our firebase. None of these rounds hit within the base.

Yet sometimes the VC were on target. I was on duty one night at the TOC when we were subjected to another heavy mortar attack. This time a VC mortar actually struck the roof of the TOC, which was a Vietnamese cement building that we had commandeered. I think it might have been a church at some time, because there were painted wooden statues of large birds stored away in a back room. The Vietnamese had an ancient animistic religion that worshiped the spirits in animals, especially birds. We had taken over the building and reinforced the roof with layers of metal sheeting and sand bags. The reinforcement worked, because that night when the mortar hit the roof, the only damage it caused was lots of dirt falling from the ceiling into our radio rooms. The roof and the building were basically undamaged. Even though I got dirty from the attack, it reassured me and made me confident in the security of the TOC.

The most frightening mortar attack on our firebase occurred on the night of December 16. By that time I was only 43 days away from going back to the world. It turned out to be the last mortar attack I had to withstand as the communists tried to kill me one more time. The enemy's mortar fire was especially damaging that night. Their mortars struck the motor pool and ignited the gasoline stores in the fuel dump. The resulting blaze quickly increased in intensity and size. The fire was spreading rapidly, which was extremely dangerous because the motor pool was located near the ammo dump where our artillery shells were stored. If the fire spread to the ammo dump and ignited the heavy artillery rounds, then everyone in the firebase would be in serious peril.

I stood with a bunch of other worried soldiers outside our bunker watching the fire. I wrote to my wife that night, "It was the most spectacular fire I have ever seen. Flames shot into the air 50 to 60 feet as the 50-gallon drums of fuel exploded. The big red cloud of fire shot upwards like an atom bomb explosion." I remember how surprised I was that those exploding fuel drums produced atomic-bomb-like mushroom clouds of flames and smoke. The fire was spreading perilously as drum after drum exploded into flames. We were all worried and talking about how bad it would be if the blazing fire reached the ammo dump.

The fire had to be contained to the motor pool. Our combat engineers were the heroes that night. Two of them jumped on a couple of bulldozers and drove them to the motor pool. I saw one of the engineers driving his bulldozer down the dirt road past our bunker. I thought, "What a courageous move," because the mortars were still falling on and near the motor pool. I had learned that moving around above ground during a mortar attack is dangerously tempting fate.

The combat engineers went to work with their bulldozers, building a dirt barrier between the motor pool and the ammunition dump. It was hazardous work, not only because the enemy's mortar shells were still landing nearby, but also because the growing fire was heating our artillery shells. If just one of those artillery shells had become hot enough to explode, it would have killed them both. Yet they kept at their work under these precarious circumstances. After working for about 20 minutes through continuing explosions of barrels of gasoline and mortar rounds, these combat engineers saved the day. They completed the construction of a dirt wall around the motor pool, which contained the huge fire. Finally the enemy's mortar fire stopped, and then other engineers, manning foam-spewing fire extinguishers, smothered all the fuel fires. We suffered no serious casualties from that potentially catastrophic attack.

The two combat engineers who worked those bulldozers were awarded the Bronze Star Medal for valor, a combat award that they rightly deserved. Their bravery during that nighttime mortar attack prevented many deaths.

30

Media Biases

The American Vietnam War attracted more intensive media coverage than any other war up to that time. It is considered the first television war. For many years, the war was the number one topic of all three major television networks on their evening news shows. Every night all across America, families watched with dread as the newscasters reported the rising body count of America's boys dying in Vietnam. Today, with our proliferation of media sources and cable channels, we must remember that the television networks' nightly news programs were the major source of information on current events for Americans during the Vietnam War. Because of this fact, network anchormen like Walter Cronkite became powerful molders of public opinion. When Walter concluded after the communists' 1968 Tet Offensive that our Vietnam War was unwinnable, citizens in middle America followed his lead.

Newspaper reporters were numerous, but many journalists covered the Vietnam War by staying in Saigon and its surroundings. This strategy ensured that they would always be near their beds, women and drugs. But of course, this restriction provided them only a limited perspective on the war. How could they understand what was really happening in the war without traveling to the rest of the country?

Television reporters and cameramen did travel throughout South Vietnam, but they were not embedded into military units as they are in our current Middle East wars. Their pattern of investigation usually went like this: in the morning the television reporter and his cameraman would catch a ride on a helicopter to a hot war zone, film and interview the commanding officer and perhaps an enlisted man or two during the day, and then fly back to Saigon before nightfall. They dreaded having to spend a night in the field because of the increased danger of nighttime, and thus avoided it at all cost. Once safely in Saigon, they would produce their story, explaining to America as an authority their interpretation of what was really happening in that war zone. But did they truly understand what was going on? Our communist enemies were secretive, their strategies devious. The war was confusing. Army commanders, who lived and fought with their units every day, would admit that they did not always fully understand what was happening. How could a civilian reporter who had spent only a few hours in the field near a battle comprehend the chaos of war?

One day, a CBS television crew followed our battalion's Charlie Company out on a combat patrol. An academy graduate, gung-ho captain commanded this frontline infantry company. His men liked to joke that their captain's principal ambition was to be Supreme Allied Commander of World War III; furthermore, they thought that his wife's primary ambition was to be the wife of the Supreme Allied Commander of World War III. Charlie Company clashed with a VC unit that morning and spent the rest of that long hot day chasing and fighting that group of VC. Charlie Company took a few casualties, but thankfully none of their soldiers were killed. However, they were never able to force the decisive battle with the VC that their company commander lusted after.

At dusk, Charlie Company formed into a protective perimeter and awaited their lift company's Hueys to fly them back to our firebase. The CBS reporter took advantage of this lull to interview the company's commander on camera. The captain stood ramrod straight as the reporter put a microphone to his face and asked questions about the day's fighting. The captain's response was basically a positive description of engaging and defeating an elusive enemy.

As the interview continued, Charlie Company's perimeter started taking incoming rifle fire. On the television broadcast, the viewer can see the reporter looking around with a worried expression as the firing starts. He continues the interview, but the incoming fire increases in volume. Now the reporter is really worried. The viewer can see infantrymen in the background anxiously diving for cover as they desperately search for the source of the fire. At this point, the television reporter is obviously frightened, because he slowly sinks to one knee. In the background, infantrymen begin to return fire. Yet the reporter continues the interview, now holding the microphone over his head as he records the captain. Throughout this time the captain ignores the danger, continues to stand at full attention, and drones on and on as the fighting around him belies his message of success.

Three days later the same Charlie Company had the bizarre experience of coming in from another long day of fighting the VC in time to watch themselves on the Armed Forces Television Network. The military channel was rebroadcasting the CBS television report of their previous clash with the VC. Everyone at our firebase was curious, and many eagerly watched the broadcast. I imagine that soldiers, sailors and Marines all over South Vietnam also watched the show, since it was the only television channel we could receive. The story highlighted that day's fighting, ending with the captain's interview as described.

I was told that the men of Charlie Company broke out into laughter at their televised captain because he looked so foolish. The captain was not watching with his men, but my informant reported that the captain was furious over the CBS broadcast. He recognized how unwise he looked, standing there at attention, claiming operational success while a firefight erupted around him. The men in his company joked that what made him so angry was that this unflattering portrayal hurt his chances of being appointed Supreme Allied Commander of World War III.

Over the course of my Vietnam tour, I gradually developed the same negative attitude toward the war's correspondents that the infantrymen in the field had. The grunts I talked to looked down upon correspondents with distain. An unbridgeable divide existed between the grunts and the reporters. To the grunts, the correspondents were like the parasitic jungle leeches they were forced to endure; both creatures were thriving on the infantryman's blood. The grunts labeled Vietnam War correspondents "combat junkies" or "war junkies," addicts who received a vicarious thrill from being close to the exciting action of combat without being exposed to its many deadly hazards. Their negative attitude also grew from the grunts' sense that they were forced to endure the blood, the mud and the death of Vietnam. It was bad enough that they had to be there; the war correspondents *wanted* to be there. How could such fools be tolerated? And this was before I saw the terribly false portrait the media broadcast of the average American soldier fighting in Vietnam, that of the "drug-crazed baby killer."

I know how deep that unbridgeable divide was between the average infantryman and everyone else, because I had experienced it once. Early in my combat tour I had been ordered out to the field to act as the battalion's RTO in one of our encirclement donut operations. Because the enemy's location was near enough to our firebase, we reinforcements were transported near to the battle by truck.

I found myself riding uncomfortably on the back of a deuce-and-one-half truck with a platoon of infantrymen. Across from me, a dirty grunt in his browned fatigues took an instant dislike to me, sitting there anxiously in my still-bright-green fatigues with my gear in a rubber bag instead of their standard rucksacks. He yelled out something obscene about me to his mates. I do not recall exactly what he said, but I do remember feeling his hatred directed at me. He was angry with me for being there. I was the *other*. Staring at his well-used M-16, I kept my mouth shut. I recall thinking that I was going through the same crap as he was, but I had better not respond antagonistically because this angry grunt might shoot me. The other men in his platoon ignored his comments.

After we dismounted from the trucks, I painfully experienced a little of what an infantryman had to endure. I had to carry my rifle and lug my bag full of resupplies as we slowly marched about two miles through the mud and the slippery rice paddy dikes in the oppressive heat and humidity of the Mekong Delta to the battle. It was not pleasant.

31

Coming Home

Time crawled slowly for me in Vietnam. I came to value routine. I wanted each day to be the same: duty at the TOC, sleeping, eating, writing letters and a little relaxation, then repeat, repeat and repeat. Breaks in the routine meant firefights, more casualties and less sleep. Yet inevitably time passed slowly, to the point that I was "short," which was our term for a man who had only a short time left in country. I was approaching the most important day for every American soldier in Vietnam, his Date Estimated Return from Over Seas (DEROS). My DEROS date was January 29, 1970, the day I would be returning to the world.

When did a Vietnam War soldier turn into a short timer? There was no official military definition; instead, there was a universally agreed upon informal period, when the soldier had less than 100 days to go in his tour of duty. This meant that he was a "two-digit midget." These lucky men could be identified by their worn out brownish-green fatigue uniforms. There was a status associated with being short; the other soldiers who were not of the class would be jealous of the short timer. In a sense, his life was valued more than that of a newbie. Everyone felt it was a greater tragedy when a short timer was killed in battle than a newbie. His friends' grief over the loss always included this sort of sentiment: "And he only had one more week to go before DEROS."

The short timer was eligible to start his short-timer calendar, which was an outlined drawing of something divided into one hundred sections. The lucky man would then fill in one section for every day that passed until the entire drawing was fully colored, sort of like an Advent calendar anticipating Christmas. DEROS was our Christmas, New Year's and birthday rolled into one. There was a wide range of styles of short-timer calendars. By far the most popular was outlined naked women, where the numbers of the days would decline closer and closer to her vagina. Other

Me in the Tan Tru mud near the end of my combat tour.

examples of short-timer calendars were of the "Freedom Bird" airplane that would fly them home or some type of appealing saying. A friend made me mine as a gift. It was a page filled with an artistic calligraphy that declared, "Candy is Dandy, but Pot Will Blow Your Mind." I appreciated his thoughtful gift. I hung it up on my cubicle/room wall and dutifully colored in every one of the hundred sections as the days passed by. It is a valuable memento of a rough time that I still possess.

There were other indicators that a soldier was short. A few carried a wooden scepter of some type; it was their "short timer's stick." Some would carve various totems onto their scepter or simply decorative designs. The scepter was a symbol of the soldier's higher status and enhanced dignity. Most short timers developed clever descriptions of their status to harass newbies with, such as, "I am so short that I have to look up to see a snake," and one of my favorites, " I am so short that I don't have time for a long talk."

In our battalion, one major advantage of being short was that troopers who had been conscientious soldiers were removed from front line positions and reassigned to safer jobs back at the firebase. A common switch was for a good soldier to be detached from a maneuver infantry company to be reassigned to our firebase's mortar platoon or to manning a radar station. These reassignments were not the result of any formal Army regulation, but rather our commanding colonel's informal reward for enduring and surviving the hazardous duty of a front line infantryman. These safer reassignments usually occurred when the dutiful soldier had completed ten months of his one-year Vietnam War tour.

My sniper roommate was a conscientious soldier; he endured extremely dangerous duty by repeatedly going out with the ARVNS on night ambush patrols. As I previously stated, the ARVN soldier was not renowned for his dedication to military discipline. My roommate did not trust the South Vietnamese soldiers he was forced to patrol with. Consequently, on night patrols he would stay awake and vigilant the entire night. After ten months of this frightening and draining duty, he thought that he deserved to be reassigned to a safer job. I agreed with his assessment. His emaciated body displayed the results of the accumulated stressors; he looked exhausted. But his commanders felt that snipers were too well trained and too valuable to be reassigned from their front line duties.

These opposing views lead to a confrontation between my sniper roommate and his first sergeant over the reassignment issue. Its climax was my roommate handing his sniper rifle to the towering first sergeant while shouting, "I am not going to kill for you anymore." Then he disappeared. He did not return to his bunk, hiding out somewhere for about two weeks. His sergeant approached me to demand that I tell him where my roommate was hiding, but I really did not know. My theory was that he had left our firebase and was moving around the Delta, hiding in other Army bases. I was wrong. Hiding out in the bush for those many months had enabled him to develop evasion skills. Later, he revealed that he had hidden out on base. His commanders eventually realized that this was a case of a good soldier who had been stressed out beyond his limits. They put out the word that if he quit hiding, he would be assigned to a safer job in radar. So he came out of hiding and his commanders were true to their word.

As a liaison specialist I already had a relatively safe job. When I became short, the only concession to my new status was that during my last few weeks, I was no longer required to go out in the field to act as an RTO on donut operations. Since these last weeks took place during President Nixon's Vietnamization strategy, I do not recall our battalion being involved in any such encirclement battles during my short phase. The ARVN colonel who commanded our joint operation South Vietnamese battalion always found ways to avoid clashes with communist enemy units. It was clear to me that he simply did not have the heart to fight. At the time, I predicted that Nixon's strategy of turning the war over to the South Vietnamese was doomed to failure. The ARVN military was no match for the motivated and disciplined NVA. Later in 1975 the rapid collapse of the South Vietnamese Republic's military proved me right.

In the Vietnam War, short timers earned status; "long timers" looked at them with respect and jealousy. We were easily identified by our worn, brown-colored fatigues, as opposed to the bright green colored fatigues of the "newbies" who had just begun their Vietnam tour. I developed a more assertive attitude and a growing confidence that I would survive this ugly war.

During this time, a new ROTC artillery second lieutenant was assigned to our firebase. He was the officer who had busted a group of soldiers smoking pot at one of our perimeter guard bunkers. (See Chapter 21 for details of that night.) After a couple of weeks of acclimation, he was assigned to work as an infantry company's FO. Veteran forward observers in the field utilized laminated maps of our AO, which could be written on with a grease pen that could then be easily erased. This new second lieutenant wanted a laminated map, so he gathered together the maps and the laminate and then carried them to the TOC one evening. I was just starting my evening duty when he came in and ordered me to make him one of those laminated AO maps. Before I could respond, he threw all his maps and laminate material at me. The supplies scattered out all over the floor.

The next morning he came into the TOC. When he saw that all the material he had thrown at me was still lying on the floor, his face fell. At that point in my Vietnam War tour, I simply could not tolerate the disrespect he had shown to me by throwing the materials at my feet, as if I was some ROTC freshman back on campus. I had earned some respect. So I stood up, looked him straight in his eyes and told him, "The United States Army does not issue second lieutenants slaves. Now if you want me to show you how to laminate a map, I will." Without saying a word he sheepishly picked up his mess and retreated. Afterward, the other enlisted men in the TOC applauded my assertive speech.

A secondary concession to my exalted short-timer status was that I was allowed to apply for an R&R leave near the end of my tour. GIs could apply for their pick of the numerous R&R sites spread across the Asian South Pacific region. I applied for Hong Kong, China and Sidney, Australia. After a couple of weeks of excited anticipation, my orders for an R&R in Hong Kong came through. Although I would have preferred Sydney, I immediately packed my duffle bag while thinking, "Is America a great country or what?" When I returned from Hong Kong, a friend informed me that my orders for Sydney had come through the day after I left. Oh, well.

My journey to Hong Kong began on January 16, 1970. I hitchhiked to Saigon, smoked a joint that was dipped in opium, hallucinated on a gigantic Army truck, and threw the rest of the pack away. While flying on a commercial jet out of Tan Son Nhut, I joked with a few Australian soldiers who all seemed to be fun-loving guys. Once in Hong Kong, I was assigned to the luxurious Hong Kong Hotel in the Kowloon section, located on the mainland of China. At the hotel I met another group of Australian soldiers staying there who were out to have a good time. Our first task was to go to the retail section of the city to buy civilian clothes. I bought a couple of inexpensive outfits and splurged on a cashmere sweater.

That night we all went out to tour the nightclubs in our new civilian clothes. We started at a club just down the street from the hotel and continued exploring one nightclub after another, as we drank, danced and joked, enjoying our escape from the war. The Australians were exceptionally social; we all got along famously.

We all became more than a little tipsy as we partied late into the night. One man voiced his concern about how we were going to find our way back to our hotel, since none of us had paid much attention to the direction we had traveled. I was a little worried because we had been warned about kidnappings of lost soldiers in the city, but when we stepped outside I laughed with relief, because I saw our towering hotel in the distance. The nightclubs were so closely packed in Hong Kong that we had only travelled three blocks from the hotel.

Hong Kong is renowned as a shopper's paradise. I did my best to take

31. Coming Home

advantage of the opportunity. Over the course of my five day R&R, I ordered custom-tailored pants, tailored shirts, a cashmere overcoat that I still own, a corduroy sport jacket and matching vest that the tailor copied from a fashion magazine picture, and the only pair of custom-made shoes I have ever owned. After the tailor measured me, the clothes were ready in three days. I spent most of my money on a strand of Mikimoto pearls for my wife and a 35-millimeter camera for myself.

I toured Hong Kong mainly by the inexpensive taxis, but one time I splurged on a rickshaw pulled by a sinewy Asian man. His thin physique belied his impressive strength. It was a unique cultural experience, yet I felt a little uncomfortable, so I tipped the man well. One day I found myself in the communist section of Kowloon, where three-story red-on-black velvet facial portraits of Chairman Mao and Lenin hung from the outsides of buildings. A large bank in this section was decorated with the usual large shiny brass entrance doors and handrail, but it also had two guards with double-barreled shotguns that were made of brass as bright as gold. The social highlight of my R&R was a celebratory dinner with one of my new Australian mates at the restaurant in the glamorous and elegant Peninsula Hotel. It was so chic that the staff posted a boy in a white uniform who stood near our table at attention, in order that he could immediately cater to our every desire. After dinner on their mezzanine I walked down the hotel's opulent central staircase feeling on top of the world in the knowledge that soon I would be returning home.

When I returned to the Tan Tru firebase, I had only ten days left before my DEROS. For the first three of those days I trained my liaison replacement at the TOC. With one week left, I started packing my gear and processing out of the military. First I had to clear my battery. Then the hated sergeant major ordered me get an extremely short haircut. The jerk had to assert his authority over me one last time. After saying goodbye to my bunker mates and promising to write, I was driven to Dong Tam to process out of the 9th Infantry Division. There I turned in my M-16 and cleared various offices. One of those required was the recruitment office, where a sergeant's job was to convince me to re-up, that is, to extend my military service. He took one look at my defiant face and just stamped my paperwork, yelling, "Get out of here!"

Next I was driven to Tan Son Nhut airbase on the northeast outskirts of Saigon. From the bus at the airport I spotted parked B-52 bombers that were absolutely sinister as their long wings drooped under the weight of their multiple jet engines. They looked like black Dragons of War. Our part of the Vietnam War ended only after President Nixon finally ordered

B-52 bombers to bomb Hanoi in 1972. When bombs started falling on the communist leaders, they decided to negotiate. Incredibly, upgraded B-52 models are still active in our present day Air Force.

At the airbase I and the other short timers awaited our Freedom Birds that would fly us back to the world. We were housed in barracks near the runway, where every morning we would assemble to hear an Air Force sergeant call out the names of the men who would be flying home that day. After that formation we were free to enjoy the many comforts of the airbase. I especially enjoyed their mess hall.

After two nights at the airbase, my name was one of those called. That day I boarded an American commercial jet airliner, my Freedom Bird. In a few minutes the pilot shot the plane high into the sky to avoid ground fire; at last I was flying back to the world. Surprisingly, there was not much cheering among my fellow military passengers, only some mild clapping. My impression was that most of us were a little dazed that the flight home was really happening. We flew to Hawaii, where we refueled. Even though someone over the speakers told us not to leave the plane, I left to find a liquor store in the airport. I quickly bought as many bottles of whiskey and rum as I could carry back to the plane. Once we were again airborne, I passed out drinks to all the men sitting near me. We celebrated going home.

After a long flight we landed at Fort Dix, New Jersey, during the cold of winter. At the fort we were issued brand new dress uniforms to wear immediately, even though we had not had a chance to shower for over two days. Such is Army logic. We were also issued winter wear, including long Army overcoats. As I was going through the line for the warm overcoats, the man next to me started to refuse his because he was going to the Deep South. I scolded him, telling him to take it anyway and then give it to a poor street person. He took the overcoat. Before we could clear the fort we were required to eat the steak dinner some woman's group had prepared for returning soldiers. So I ate a steak that I really did not want; I just wanted to get out of the military. I returned from the Vietnam War and was honorably discharged with the Good Conduct, Vietnam Service, Vietnam Campaign and Army Commendation Medals that same day: February 3, 1970.

Free at last! I took a taxi to the Philadelphia airport. While at the airport, I sold the rest of my bottles of liquor to an attendant for five dollars because I had too much to carry. I flew on a commercial airliner to Pittsburgh. We had been warned that there might be anti-war protestors at airports, but there were none in the two biggest Pennsylvania airports I passed through. Once I arrived at the Pittsburgh airport, there were no brass bands either, or any other type of community welcome home. My impression of the civilians was one of indifference: "Oh, you're home from the Vietnam War. Ho hum, I have to get to work." Of course, my wife met me at the airport and we had a joyful reunion. She drove me back to our new apartment. The next day, to celebrate, we drove to the Shadyside section of Pittsburgh for lunch. This section of the city is a hip district near the campuses of the University of Pittsburgh and Carnegie Mellon University. As we were having lunch, I spotted a good female friend from my high school. Happy to see her, I ran up to her, shouting, "I'm home." Her response: "Where did you go?"

32

Reflections

Resurrecting my remote memories of the Vietnam War has been a challenging endeavor. At times, the project put me in a bad mood. The most upsetting experience was reading through my many letters to my wife. The haunting pangs of my loneliness and depression spring out from the letters. Yet writing my war memoir has proven to be a productive form of self-analysis. Besides recognizing that Vietnam was the tempestuous adventure of my lifetime, the recollecting has forced me to admit that my Vietnam War tour was the major turning point of my life. If I had not gone to war, I could still be teaching psychology in a Pittsburgh suburban high school. Fighting in the Vietnam War is an experience I would not repeat for millions of dollars, yet at the same time I am proud that I served my country and survived.

As the philosopher Friedrich Nietzsche wrote, "That which does not kill us, makes us stronger." Directly facing the prospect of death made me a stronger man. I was already a college graduate, but the Vietnam War is where I received my real education. It taught me the reality of my mortality and a deeper appreciation of life. My horizons widened with new possibilities because of my war experiences. The Vietnam War taught me to discard a narrow-minded religion and replace it with a new worldview, an international citizen-of-the-world perspective. I am a more complex person than I would have been otherwise. Forcing me to relate to a wide range of human types enhanced my personal development. Also, I learned how to be comfortable and sleep anywhere.

The Vietnam War hastened my marriage to a wonderfully talented and loving woman, Judith Anne, who has stuck by me for these many decades. We had an incredibly romantic wedding and honeymoon in Hawaii that would never have occurred otherwise. Together we raised an intelligent, considerate and attractive daughter, Colleen Michele. Last year she and her husband, Dr. Adam November, blessed us with a delightful grandson, Finley Daniel. These four are my balcony people, who watch over me and cheer me on. They are the joys of my life.

Combining Vietnam with my second most influential life experience, my graduate studies in clinical psychology at Indiana University, enabled me to turn the sour lemon of Vietnam into the sweet lemonade of helping Vietnam combat veterans recover from their psychological war wounds. The chief of psychology at the Veterans Hospital in Milwaukee, Dr. Kenneth Klauck, hired me to lead a Vietnam veteran counseling program, the Vet Center, because of my Vietnam War experience as well as my doctorate in clinical psychology.

Fighting in that horrible war enabled me to communicate with troubled veterans in an empathic manner that the veterans always appreciated. I feel a bond with other combat veterans, an instantaneous respect for the hell they have gone through. We are a band of brothers. Helping hundreds and hundreds of Vietnam War veterans emotionally return home from the war was more than a job; it was a crusade. I am proud of my early advocacy for and my expertise in post-traumatic stress disorder (PTSD), as well as my advocacy for combat veterans of all of America's recent wars. This effort culminated with the publication in 2012 of my book, *Understanding Combat-Related Post-Traumatic Stress Disorder*, which has been added to college libraries across America and in many foreign countries.

But the aftereffects of the war on me were not all positive. It vividly demonstrated to me that real evil exists in the world. It made me always afraid of the bad that could happen. For the longest time my daughter thought that I was a total pessimist, until I explained to her to her that I had learned in the Vietnam War to predict the worst that could happen. Then I did everything I could to prevent it or to minimize its negative effects. I explained that that is why, for example, I have bought so many different types of insurance policies, or why I buy cars with all the latest safety features. I have become risk averse because I can too easily imagine the negative outcome possibilities.

My self-analysis as a result of writing this memoir revealed another change in my psychological characteristics due to the war. I no longer have the capacity to "forgive and forget." I cannot forgive the Vietnamese communists for the despicable inhuman violence they displayed against American and ARVN soldiers, as well as toward Vietnamese civilians. I will not forget the thousands of my brother soldiers that the communists killed in horrible ways, or the hundreds of thousands they wounded. I cannot forgive the communists for traumatizing American soldiers so badly that many have not been able to live normal lives afterward. I will not return to Vietnam, even though I have been offered many chances to do so. I did not want to go there the first time; why should I go back? In my current life, if you hurt me, I do not forgive and forget. I do not actively seek revenge, but I remember the pain.

The Vietnam War instilled in me a capacity for rage that frightens me a bit. While in graduate school during those first years back from the war, I lost my temper a couple of times, and it was never pretty. Fortunately, there were no serious consequences of my screaming episodes, but they demonstrated to me that my anger could be difficult to control. Consequently, I joined Indiana University's Karate Club for three years. The practice improved my self-confidence and self-control. Yet even today some observer may see me in a situation where it appears that I am afraid of someone or something. The internal reality is that I am not afraid of the person, but I am afraid that if I confront them and they resist, I may become enraged. So I tend to avoid direct confrontations.

I am afraid of everything and afraid of nothing at the same time. This is a true statement. I am an expert at being afraid because I have had extensive practice. I am afraid of everything bad that could happen. But at the same time I possess confidence in myself that I will effectively deal with whatever problem occurs. I have proven to myself in Vietnam and in difficult situations afterward that I can cope with stressful situations. This self-confidence is one of the reasons why I was successful at treating imminently suicidal patients. Another reason is my firm belief in the sacredness of life. I have no desire to do so, yet if seriously threatened, I would fight to the death to protect my family and myself.

While she was a child, I taught my daughter the military concept of "field expediency" that I had learned in Vietnam. It is the attitude that if she were stuck somehow and did not have the correct tool to fix the problem, she should not give up. Field expediency means that she needs to be flexible, to think creatively and improvise in that problematic situation. Over the years I have demonstrated the concept by fixing problems, not with the ideal tool, but with the tools at hand. It may have been as simple as opening a wine bottle with a knife instead of a corkscrew, or as complicated as finding alternative ways of accomplishing a task on our computers. I also taught her that, at times, field expediency means that she might have to build the tool that will do the job. I have demonstrated that to her, too.

Recently, decades later, my wife and I had to evacuate our home in Florida because of a threatening hurricane. Friends asked Colleen, who lives in San Francisco, if she was worried about her parents' evacuation.

She told me later that her answer was, "No." She told her friends that her father knew and applied field expediency; therefore, she was confident that I could master any challenge that might occur in the evacuation. We were lucky; the hurricane did not strike land near our home.

In the 1970s, Veterans Administration Hospital psychologists and psychiatrists were treating patients with psychological symptoms related to their combat traumas in the Vietnam War. They started calling the disorder "post–Vietnam syndrome" because the previous diagnostic terms of shell shock or combat fatigue did not fit these new veterans. Previously, psychological symptoms due to combat were thought to be temporary in nature, but now the doctors were treating Vietnam War veterans whose symptoms surfaced years after the war.

As stated, most of the young American soldiers in Vietnam did not really understand communism or the reasons why America was at war in this primitive, far-off country. To give us an honorable reason for fighting, some positive motivation to hold on to and to look forward to, every soldier I knew romanticized "back home." Every soldier's hometown was the best place to live in America; every soldier's girlfriend was the best looking girl in his high school. We had to survive this gruesome war so we could return to the best hometown and marry the prettiest girl. Each of us idealized our part of America.

None of us had the slightest idea of the false negative picture of Vietnam veterans that the media had planted in the minds of our peers. None of us predicted the rejection we would receive from our America. We did not expect celebratory brass bands, but no one expected attacks. We are not the baby killers Americans thought we were. The American people did not want to accept the true reason for that war, that they had elected Democratic presidents who escalated a regional conflict into the decade-long Vietnam War that could have triggered World War III. The American people did not want to accept their responsibility for the war; they needed a scapegoat to blame it on. Disoriented upon returning, we were easy targets. With the media's encouragement, Americans blamed the war on the warrior. Their argument was, "He pulled the trigger." But who put the rifle in the soldier's hands? Who put that young man in the middle of a kill-or-be-killed war zone? It is analogous to blaming a rape on the victimized woman.

The persistence and severity of Vietnam War veterans' psychological problems is partially due to the horrible reception we received upon returning to the United States. We were America's scapegoats. Many Vietnam veterans I treated were met with violence at the airports. They told me of war protestors spitting on them and throwing bags of urine and feces at them. The Vietnam veteran's homecoming was like that of a woman who returns from the hospital after being brutally raped to a husband who views her as damaged goods, morally tainted and impure. He may even blame her for wearing that short skirt or for going out shopping that night, as if she wore the skirt to attract a rapist. Sadly, many times the confused rape victim comes to believe the distortions and begins, in fact, to blame herself: "Why did I wear that short skirt that night?"

We can understand how this dense husband's non-empathetic reception will aggravate the rape victim's symptoms and make her recovery more difficult. What she needs is emotional support and understanding from those who care for her. It is exactly the same for the American Vietnam War veteran. Being rejected by his friends, his community, and, most tragically, even by his own family aggravated all veterans' war-caused psychological problems. Those multiple rejections made each veteran's emotional recovery much more difficult. Many came to believe the "big lie" that the media broadcasted, causing them to suffer concomitant guilt feelings. Sadly, many Vietnam War veterans never did recover. Too many of these poor souls committed suicide, either actively or passively through alcoholism. Others are still out there, living lonely, isolated lives.

No other group of combat veterans from America's other wars suffered such rejection from the very country that had sent them off to war. The popular media and most anti-war protestors lied. Vietnam War veterans were not the cause of that horrible war. The vast majority of Vietnam War veterans did their duty and served their country honorably. They did not commit atrocities and did not deserve the attacks. Readers, if you know of, work with, or live near a Vietnam War veteran, thank him for serving our country. Perhaps you could give him a hug too. He deserves your appreciation; he needs it.

Vietnam Glossary

AK-47	Automatic rifle of communist soldiers
AWOL	Absent With Out Leave
AIT	Advanced Individual Training
AO	Area of Operations
ARVN	Army of the Republic of Viet Nam
ASAP	As Soon As Possible
BAR	Browning Automatic Rifle
C-Rats	Canned Rations
CBS	Columbia Broadcasting System
C&C	Command & Control
CO	Commanding Officer
DEROS	Date Expected Return from Overseas Service
DI	Drill Instructor
EM	Enlisted Man
FDC	Fire Direction & Control
FO	Forward Observer
FM	Frequency Modulation
GI	Government Issue … evolved into a nickname for American soldiers
H&I	Harassment & Interdiction
ID	Identification
JAG	Judge Advocate General, the term for military lawyers
KIA	Killed In Action
KP	Kitchen Police
LZ	Landing Zone
LOH	Light Observation Helicopter
M-16	Automatic rifle of American and allied soldiers

MACV	Military Assistance Command Vietnam
MOS	Military Occupation Specialty
MPC	Military Payment Certificates
MP	Military Police
NCO	Non-Commissioned Officer
NVA	North Vietnamese Army
OCS	Officer Candidate School
OJ	Opiated Joint
PX	Post eXchange, military retail store
PTSD	Post-Traumatic Stress Disorder
RTO	Radio Telephone Operator
RECON	Reconnaissance platoon
ROTC	Reserve Officer Training Corps
R&R	Rest & Recreation
RPG	Rocket-Propelled Grenade
TOC	Tactical Operations Center
TOT	Time On Target
USO	United Service Organization
VC	Viet Cong, communist guerrilla fighters
WIA	Wounded In Action

Military Service History of Walter F. McDermott

Volunteered for the United States Army in Pittsburgh, Pennsylvania, on June 25, 1968

- Basic Training at Fort Leonard Wood, Missouri
- Advanced Individual Training at Fort Sill, Oklahoma
- Military Occupation Specialty: 13E20, Artillery Fire Direction & Control

Ordered to the Republic of South Vietnam on January 2, 1969

- Assigned to the 2nd of the 4th Artillery attached to the 9th Infantry Division
- Stationed at the headquarters of the 2nd of the 4th Artillery Brigade at Tan An, Mekong Delta, South Vietnam
- Worked as a radio telephone operator, promoted to Specialist Fourth Class (E-4), March 1, 1969

Transferred to the 2nd of the 60th Infantry Battalion of the 9th Infantry Division stationed at Tan Tru, Mekong Delta, South Vietnam, at the end of March 1969

- Assigned to the 2nd of the 60th's Headquarters Company
- Worked as a liaison specialist, coordinating artillery fire for the infantry

Departed South Vietnam on DEROS date: January 29, 1970

- Awarded Army Commendation, Vietnam Campaign, Vietnam Service, National Defense and Good Conduct medals
- Discharged from active duty at Fort Dix, New Jersey, on February 3, 1970
- Discharged from the inactive reserves on June 24, 1974

Index

AC-130 Spector gunship 119
AK-47 rifle 14, 39, 115–117, 146, 168
Apollo 11 rocket 129
Armstrong, Neil 129
Army Commendation Medal 41, 198

Baumgarten, Harold 171
Blackstone Rangers 12
Bronze Star Medal 41, 187
Brown Water Navy 46, 91, 92

C-rations 112- 114
Chinese 107-mm. rockets 69–72
Cobra Gunship 40, 60, 64
conscientious objector 16, 167
Cronkite, Walter 188

Dong Tam 77, 90, 98, 128 168, 174, 179, 197
Douglas A-1 Skyraiders 124
Douglas AC-47 Transport 118, 119
drill instructor 13- 17, 19, 25, 29
Dust Off medical helicopter 7

F-4 Phantom Jets 51- 53
field expediency 202
Fort DeRussy 177
Fort Leonard Wood 11, 15
Fort Sill 21, 22
Franklin, Aretha 153
Freud, Sigmund 18

Gatling, Richard 118
General Giap 103
Golan Heights 33
Green Berets 1, 179

Henry IV, Part 1 2
Ho Chi Minh 37, 56, 103, 171
Ho Chi Minh Trail 37, 73
Hong Kong 196
Howitzers 23, 24
Huey helicopter 5, 59, 65–67

Immaculate Heart of Mary Church 165
Indiana University, Bloomington, Indiana 19, 81, 201, 202
Indiana University of Pennsylvania 10

Johnson, Lyndon 89

Kissinger, Henry 149
Klauck, Kenneth 201
Kristofferson, Kris 106, 140
Kubrick, Stanley 26
Kulakowski, Frank 32
Kulakowski, Joseph 31
Kulakowski, Mary 31

Lawton, Oklahoma 22
light observation helicopter 40, 62–65

M-14 rifle 13, 19, 116, 133, 148
M-16 rifle 4, 102, 115–117, 125, 126, 160, 162–164, 168
M-60 machine gun 5, 40
M76 fragmentation grenade 100
M-134 mini guns 118, 119
McDermott, Judith 174, 176, 177, 200
McDermott, Walter, Sr. 32
McNamara, Robert 103
Mekong Delta 4, 5, 35, 79, 86, 88, 92, 191

Napalm 52
Nietzsche, Friedrich 200
Nixon, Richard 149, 151, 152, 195, 197
North Vietnamese Army 67, 94, 103, 120, 125, 126, 141, 146–148
November, Adam 201
November, Colleen 200, 202, 203
November, Finley 201

Oakland, California 32
officer's candidate school 11- 13, 18, 19, 22, 29, 30
Oklahoma State University 27
105-mm. Howitzers 44–46, 48, 49, 73, 74
106-mm. recoilless rifle 183, 184
155-mm. Howitzers 44, 45, 147, 148
175-mm. Long Tom Howitzers 77

P-38 can opener 112
Piper Cub Observation Airplanes 51

Pittsburgh, Pennsylvania 10, 13, 21, 31, 122, 165, 167, 19
Plain of Reeds 73, 74, 88
Polish Hill 165
post traumatic stress disorder 201
PRC-25 radio 4, 5

rocket propelled grenades 94, 120

Saigon 33, 34, 139, 140, 154–159, 188
Saving Private Ryan 171
Scislowski, Walter 31, 32
Shakespeare, William 2
shell shock 70
South Korean Army 46
Spielberg, Steven 171
Stalin, Joseph 89, 166
Stars and Stripes (newspaper) 144, 145, 170

Tan An 35–43, 90, 131, 160–162
Tan Son Nhut Airbase 33, 34, 51, 111, 140, 197
Tan Tru 3, 8, 41, 42, 44, 169
Tet Offensive 10, 39, 188
trigonometry 23
Tu Di 84, 85
2001: A Space Odyssey 26

Viet Cong 5–7, 56, 57, 146, 170, 171, 173, 178, 183–186

Waikiki Beach, Hawaii 175–177
Wayne, John 1, 179, 180, 181
West Point Military Academy 48, 142, 143, 174

Yom Kippur War 33

www.ingramcontent.com/pod-product-compliance
Ingram Content Group UK Ltd.
Pitfield, Milton Keynes, MK11 3LW, UK
UKHW041959140426
5217IPUK00015B/883